ENVIRONMENTAL REMEDIATION TECHNOLOGIES, REGULATIONS AND SAFETY

SUSTAINABLE UTILIZATION AND MANAGEMENT OF NATURAL RESOURCES IN THE ERA OF CLIMATE CHANGE

ENVIRONMENTAL REMEDIATION TECHNOLOGIES, REGULATIONS AND SAFETY

Additional books in this series can be found on Nova's website under the Series tab.

Additional e-books in this series can be found on Nova's website under the e-book tab.

ENVIRONMENTAL REMEDIATION TECHNOLOGIES, REGULATIONS AND SAFETY

SUSTAINABLE UTILIZATION AND MANAGEMENT OF NATURAL RESOURCES IN THE ERA OF CLIMATE CHANGE

EDWARD DEBRAH WIAFE
AND
FRANK S. ARKU
EDITORS

nova publishers
New York

Library of Congress Cataloging-in-Publication Data

Sustainable utilization and management of natural resources in the era of climate change / editors, Edward Debrah Wiafe and Frank S. Arku (Department of Environmental and Natural Resources Management, Presbyterian University College, Akropong-Akuapem, Ghana).
pages cm. -- (Environmental remediation technologies, regulations, and safety)
Includes index.
ISBN 978-1-63482-896-3 (hardcover)
1. Nature conservation--Ghana. 2. Natural resources--Ghana--Management. I. Wiafe, Edward Debrah. II. Arku, Frank S.
QH77.G43S87 2015
333.7209667--dc23
2015016700

Published by Nova Science Publishers, Inc. † New York

CONTENTS

PREFACE

Over the past millennia, natural resource utilization and management on a sustainable basis has been the issue of many field studies. There have been concerns of declining natural resources and many reasons have been attributed to this pandemic. Efforts to address the situation have now become an integral part of governments' policies and programmes.

The social and economic developments are occurring at an accelerated rate at the expense of natural resources. Over the last decade, there has been a growing interest in natural resources and their utilization to meet the human needs that has been driven by concerns about global climate change. Other developmental projects are the construction industry, where uncultivated lands have given way to estate development.

This volume includes chapters by biological and sanitation scientists trying to achieve a balanced viewpoint on how natural resources can be sustainably utilized in the dispensation of climate change.

Chapter One deals with the need to conserve carnivores in protected areas and echo the efforts to restore their populations.

Chapter Two describes the density and distribution of birds in an ecotourism site. It shows the factors that affect the avian population and how birds have positively impacted on tourism.

Chapter Three deals with community based natural resources and their efforts to assess the growing tree stock for sustainable yield allocation in the community. There are heterogeneous groups of resources in nature which are difficult to quantify and Chapter Four demonstrates how these complex, non-timber forest products can be inventoried to inform decision making on its utilization. Solid waste management is a current topical issue confronting politicians and natural resource managers.

Chapter Five investigates solid waste composition and its recovery in relation to seasonal changes.

Chapter Six reviews empirical evidence on sustainable energy utilization in the era of climate change.

Finally, it is our hope that this book will bring the fundamental dilemma of utilization and sustainability to public attention, start the debate and move a step further toward natural resource conservation.

CONTRIBUTORS

Emmanuel Danquah
Department of Wildlife and Range Management,
Faculty of Renewable Natural Resources,
Kwame Nkrumah University of Science and Technology,
PMB, Kumasi, Ghana.

Emmanuel Danquah and Evans Atsu Abormega
Department of Wildlife and Range Management,
Faculty of Renewable Natural Resources,
Kwame Nkrumah University of Science and Technology,
PMB, Kumasi, Ghana.

Edward D. Wiafe
Department of Environmental and Natural Resources,
Presbyterian University College, Ghana,
P. O. Box 393, Akropong Akuapem, Ghana.

Francis Kofi Bih
College of Technology Education,
University of Education, Winneba,
P. O. Box 1277, Kumasi, Ghana.

Richard Amfo-Otu, John Adu Kumi and Foster Kwakye
Department of Environmental and Natural Resources Management,
Presbyterian University College Ghana, Akuapem Campus,
P.O. Box 393, Akropong-Akuapem,
Email: richard.amfo-otu@presbyuniversity.edu.gh

Department of Agribusiness, Presbyterian University College Ghana, Akuapem Campus, P.O. Box 393, Akropong-Akuapem,
Email: adu-kumi.j@presbyuniversity.edu.gh
Department of, Accra polytechnic,
P. O. Box, Email: foster2simple@yahoo.com

Paul Adjei Kwakwa, Solomon Aboagye and Hamdiyah Alhassan

Presbyterian University College Ghana, Okwahu Campus, Ghana
Corresponding author E-mail: pauladkwa@yahoo.com
Department of Economics, University of Ghana, Legon – Accra, Ghana
E-mail: sbaboagye@gmail.com
Department of Agricultural and Resource Economics, University for Development Studies, Nyankpala Campus, Ghana
Email: abena567@yahoo.co.uk

In: Sustainable Utilization
Editors: E. D. Wiafe and F. S. Arku
ISBN: 978-1-63482-896-3

Chapter 1

IMPORTANCE OF PROTECTED AREAS FOR CARNIVORE CONSERVATION IN GHANA

Emmanuel Danquah*
Department of Wildlife and Range Management,
Faculty of Renewable Natural Resources,
Kwame Nkrumah University of Science and Technology,
PMB, Kumasi, Ghana

ABSTRACT

In the current era of global climate change, protected areas are vital to national biodiversity conservation strategies. In order to determine the relative importance of categories of protected areas to carnivore conservation, species composition and encounter rates of carnivores were examined in two different protected area categories in the Bia-Goaso Forest Block in western Ghana. One hundred and ten line transects were systematically distributed in an extensive network of two wildlife reserves and nine forest reserves. Carnivore sightings and signs (droppings and tracks) belonging to six species were recorded during the survey. The most abundant species were blotched genet (*Genettatigrinapardina*; 25.1% of signs) and slender mongoose (*Herpestessanguineas*;23.1%). Others were marsh mongoose (*Atilax-paludinosis*; 18.7%), African civet (*Civettictiscivetta*; 18.0%) and

---

* Email: emmanueldanquah@yahoo.com, emmanueldanquah@gmail.com; Phone: +233-244 742 385.

Cusimanse mongoose (*Crossarchusobscurus*; 12.2%). The only large carnivorewas the leopard (*Panthera pardus*) and had a relative abundance of less than 1.0%. Both carnivore density (Mann-Whitney U test: U=1797.0, P<0.01) and diversity (U=1648.5, P<0.01) were significantly greaterin wildlife reserves than forest reserves. The findings suggest that forest reserves maybe achieving partial success in protecting wildlife, whereas wildlife reserves are considerably more effective, although not entirely successful. To reduce future negative global climate change impacts, management approaches that include more wildlife protection in forest reserves are essential in order to effectively conserve carnivores in Ghana's protected areas.

Keywords: Ghana, mammal, species, carnivores, transect, protection

1. INTRODUCTION

Mammalian carnivores are vulnerable to extinction in fragmented landscapes due to their low densities, extensive spatial requirements, inevitable conflict with people [1] and gloomy prospects of global climate change impacts. [2] Large, intact landscapes therefore remain critical to the survival of many carnivore populations. However, there is increasing evidence that even intact landscapes alone may not effectively conserve carnivore populations, especially when dealing with future negative climate change impacts. [2] Also, abundance of species differs among heavily hunted and less huntedplaces. [3, 4]To avoid the negative effects of climate change and human activities, including hunting, and to conserve natural and cultural features of the land, many protected areas have been established throughout the world. Protected areas are therefore the cornerstone of most national biodiversity conservation strategies. [4]

Unfortunately, little information exists on how protected area types influencecarnivore composition. Data on abundances of carnivores in different protected area categories and the factors that affect them are required for a number of reasons including: monitoring for future climate change impacts, setting of management priorities; measuring the effects of management; managing populations of problem carnivores; assessing the effects of human activities and providing evidence for the need for policy change. [5]

Consequently, a comparison of carnivore encounter rates and species richness in 11 reserves belonging to two protected area management categories (2 wildlife reserves and 9 forest reserves), with similar environmental characteristics but different hunting restrictions and levels of protection was conducted. The hypothesis was that encounter rates and species richness of mammals in forest reserves would be less than in wildlife reserves, where hunting is prohibited and there is better natural-resource protection through law enforcement. It is hoped that the findings will generate broader discussion and help conservationists to identify the most appropriate and effective protected area approach to manage and monitor carnivores for future climate change impacts.

2. Study Area

The Bia Goaso Forest Blockis located in Western Ghana and comprises the Bia Conservation Area (Bia National Park and Bia Resource Reserve) and an extensive network of nine Forest Reserves (Mpameso, Bia Tano, Asukese, Bonkoni, Krokosua Hills, BonsamBepo, Ayum, Subim and Bia North). The area extends from latitudes 5.20 to 6.20 degrees north and longitudes 2.20 to 3.10 degrees west, south of Sunyani to the west of the Tano River and to the Ghana-Cote d'Ivoire border(Figure 1).

The natural land cover corresponds to the Guinea-Congolian forest vegetation. [6] The vegetation is both moist and dry semi-deciduous type (Hall & Swaine, 1981). This matches with the *Celtiszenkeri-Triplochitonscleroxylon* association. [7]The mean elevation is 200–550m, with generally undulating topography. Mean annual rainfall is 680–1450 mm/year, characterized by a bi-modal wet season from March to July and September to November and a major dry season from December to February.

3. Methods

One-kilometre line transects were distributed systematically over a numbered grid at intervals of 2 km in each of the protected areas. This gave 110 transects (20 transects in wildlife reserves and 90 transects in forest reserves (Figure 1). Transect orientation was perpendicular to the main drainage lines of the region. [8]

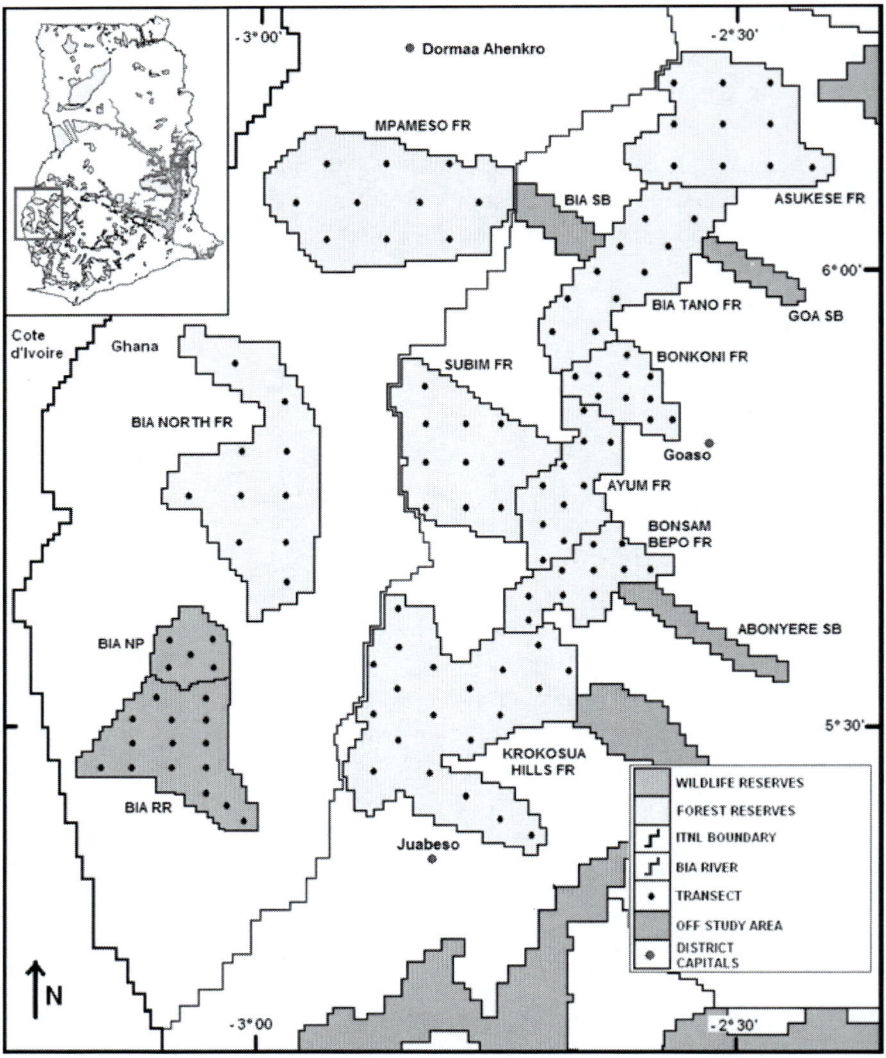

Figure 1. Study area showing distribution of transects in the wildlife and forest reserves. The inset map shows the location of the study area in south-western Ghana.

Four field teams of four persons each walked the line transects looking for carnivores and their signs (droppings and tracks). Each time a carnivore or its sign was encountered, the species and place the sign was found was recorded. The tracks of a carnivore crossing the trail was counted as one sighting. Likewise, when carnivore footprints were following the trail, they were considered as one observation. In the case of gregarious species, such as mongoose, one group of tracks was counted as one sighting.

An encounter rate index per km for each species was estimated by dividing the number of carnivore encounters (direct sightings and signs) by the total length of a given transect. A Mann-Whitney U test was used to compare the encounter rates and species richness of carnivores between wildlife and forest reserves. All analyses were conducted using the Statview software. [9]

4. RESULTS

Carnivore signs (droppings and tracks) belonging to six species were confirmed on transects during the survey period (Table 1). The most commonly registered signs were those of the blotched genet (Genettatigrinapardina) which represented 25.1% of all carnivore signs and slender mongoose (Herpestessanguineas;23.1%). Others were marsh mongoose (Atilaxpaludinosis; 18.7%), African civet (Civettictiscivetta; 18.0%) and Cusimanse mongoose (Crossarchusobscurus; 12.2%). The only large carnivore, leopard (Panthera pardus) had relative abundance of less than 1.0%.

Table 1. Summary of Carnivore Species Characteristics and Conservation Status

Functional Type	Body Mass (kg)	Common Name	Scientific Name	Conservation Status (IUCN 3.1)
Large Carnivores	20 - 90	Leopard	*Pantherapardus* (Linnaeus, 1758)	Near Threatened
Medium Carnivores	11 – 18	African Civet	*Civettictiscivetta* (Schreber, 1776)	Least Concern
	3 – 6	Marsh Mongoose	*Atilaxpaludinosus* (G.[Baron] Cuvier, 1829)	Least Concern
	1.5 – 2.6	Blotched Genet	*Genettatigrinapardina* (I. Geoffroy Saint-Hilaire, 1832)	Least Concern
	0.4 – 1	Cusimanse	*Crossarchusobscurus* (F.G. Cuvier, 1825)	Least Concern
	0.4 – 0.8	Slender Mongoose	*Herpestessanguineas* (Rüppell, 1835)	Least Concern

Both carnivore density (Mann-Whitney U test: $U=1797.0$, $P<0.01$) and diversity ($U=1648.5$, $P<0.01$) were significantly greaterin wildlife reserves than forest reserves. With the exception of leopards, none of the species were unique to any of the protected area categories (Table 2). The generally higher species density and richness in wildlife reserves in comparison to forest reserves may result from the disproportionately higher numbers of most carnivores. Among functional types, medium-sized carnivores dominated (83 percent of carnivores) in the study area whilst large carnivores constituted only 17 percent of the group surveyed.

Table 2. Summary of Carnivore Species Abundance Characteristics

			Mean Encounter Rates	
Common Names	Total Signs	Encounter Rates	Wildlife Reserves	Forest Reserves
Medium-sized Carnivores				
Slender Mongoose	163	1.5	5.5	0.6
Cusimanse Mongoose	86	0.8	3.4	0.2
Marsh Mongoose	132	1.2	4.5	0.5
Blotched Genet	177	1.6	6.1	0.6
African Civet	127	1.2	4.5	0.4
Large Carnivores				
Leopard	21	0.2	1.1	0.0
Total	706	6.4	24.9	2.3

Discussion

The abundance of carnivores in the study area varies considerably between the two major categories of protected areas. [2, 10] Carnivoreencounter rates, in forest reserves were significantly lower than those in wildlife reserves, where hunting is regulated and possibly there is lower risk to anthropogenic mortality rates. This supports the hypotheses that species richness and encounter rates of carnivores in forest reserves would be lower than in wildlife reserves, where hunting is prohibited and there is better natural-resource protection through law enforcement. Large carnivore species such as leopards might have benefitted largely from this scenario. The apparent absence of leopards in forest reserves is presumably unrelated to exploitation, because the species is not usually utilized in the area.

Management goals for protected areas, which act as reservoirs for harvesting in adjacent areas typically require populations to be at, or close to, carrying capacity. [1] This suggests that, although carnivores seem adequately protected in wildlife reserves, differential anthropogenic mortality and lower probability of survival might have kept the carnivore population well below carrying capacity in forest reserves. While it is unlikely that carnivores will disappear completely from forest reserves, given the proximity of other protected populations, the overall effectiveness of the protected area system appears weakened by higher persecution in forest reserves and associated effects on vital rates.

The most effective options for combating such effects are increasing forest buffers around protected areas or decreasing the number of people living outside its boundaries. [1, 11] However, with the rapid expansion of local human populations, both are unlikely. A suitable alternative is the implementation of conservation measures in fringe communities. [12] In 2000, the Wildlife Division policy for collaborative community based wildlife management was instituted to secure Ghana's network of protected areas by developing wildlife management as a legitimate and viable land-use option in fringe communities. [13] The strategy is to promote wildlife as a viable complement to existing agricultural practices through the sustainable production of bushmeat and other wildlife products. This involves developing a working partnership between the protected areas and people living in and around them to each other's benefit.

As human populations continue to increase around protected areas and climate change has emerged as a global problem, approaches that manage human activities on and off-reserves will be essential if conservation of wide ranging taxa is to succeed. [14] Relatively resilient, large carnivores such as leopards can act as substitutes for conservation outcomes that benefit other species. [15] At the Bia Goaso Forest Block, activities that mitigate anthropogenic activities for leopards would favour civets, genets and mongooses, although implementation would need to occur on a smaller scale to benefit civets, genets and mongooses, given their less demanding area requirements. [16] Similar non-carnivore potential surrogates include elephants and chimpanzees. Understanding the impact of anthropogenic effects on carnivore populations and devising appropriate mitigation strategies might hold promise for buffering, maintaining or restoring landscapes that are suitable for more locally endangered species such as leopards.

Conclusions and Recommendations

The study suggests that the category of protected area has an effect on the populations of carnivores. [4] The type of activities allowed in the protected areas and the level of law enforcement may be directly affecting the abundance of carnivores and mammals in general. Evidence from [2, 10] suggests that the main factor differentiating the abundance of mammals in the two protected areas categories considered could be the level of hunting. Although all carnivore species were influenced, the ones that were more severely affected were larger carnivores. Hence, forest reserves seem to be achieving only partial success in protecting wildlife, whereas wildlife reserves seem to be considerably more effective, although not entirely successful. [4]

Restricting hunting activity in the forest reserves of Ghana is not easy and perhaps impractical as long as the current socioeconomic conditions persist. Yet hunting should be sustainable so that many large animals do not become extinct in the region. [4] The goal of sustainable wildlife utilization can be reached only if we have basic information on the populations of wildlife in the area so that changes in their abundance due to negative human activities (including hunting) can be assessed and monitored. Standardization of methods to undertake these assessments in tropical forests is of foremost importance. It is also necessary to work with the communities that live in and around protected areas: if their standards of living improve, then pressure on wildlife populations will be minimized. [4]

ACKNOWLEDGMENTS

I wish to acknowledge the Ghana Forestry Commission for provision of field staff, logistics and granting permission to work within the reserves. Special thanks go to the management of the Wildlife Division and Forestry Division.

The study would not have been possible without the contributions of the field team.

REFERENCES

[1] Balme, G., A., Slotow, R. and Hunter, L., T., B. (2010). Edge effects and the impact of non-protected areas in carnivore conservation: leopards in the Phinda–Mkhuze Complex, South Africa. *Anim. Conserv.* 21, 1–9.

[2] Danquah, E., Sam, M., K. and Marshal, P. (2009a).*Monitoring Matters: A survey of large mammals of the Ankasa and Bia Conservation Areas.* Report submitted to EU/WD Protected Areas Development Programme Phase II Project.

[3] Lopes, M., A. and Ferrari, S., F. (2000). Effects of human colonization on the abundance and diversity of mammals in eastern Brazilian Amazonia. *Conserv. Biol.* 14, 1658-1665.

[4] Carrillo, E., Wong, G. and Cuarón, A., D. (2000). Monitoring mammal populations in Costa Rican protected areas under different hunting restrictions. *Conserv. Biol.*14, 1580-159.

[5] Battersby, J., E. and Greenwood, J., D. (2004).Monitoring terrestrial mammals in the UK: past, present and future, using lessons from the bird world. *Mammal Rev.* 34: 3-29.

[6] Hall, J., B. and Swaine, M., D. (1981). *Distribution and ecology of vascular plants in tropical rainforest forest: forest vegetation in Ghana.* Junk, The Hague.

[7] Taylor, C., J. (1960). *Synecology and sylviculture in Ghana.* Thomas Nelson and Sons, Edinburgh.

[8] Norton-Griffiths, M. (1978). *Counting animals.* African Wildlife Foundation, Nairobi.

[9] SAS Institute. (1999) *Statview for windows version 5.1.* Cary NC, USA.

[10] Danquah, E., Sam. M., K., Akom, E., Appiah-Kubi, S., K. and Ayiku, T. (2009b).*Elephant Conservation and Possibilities of Creating Corridors in Western Ghana.Project_AFE-0348.* Technical report submitted by A Rocha Ghana to African Elephant Conservation Fund of the US Fish and Wildlife Service.

[11] Harcourt, A., H., Parks, S., A. and Woodroffe, R. (2001). Small reserves face a double jeopardy: small size and high surrounding human density. *Biodiv. Conserv.* 10, 1011–1026.

[12] Forbes, G., J. and Theberge, J., B. (1996). Cross-boundary management of Algonquin pack wolves. *Conserv. Biol.* 10, 1091–1097.

[13] Wildlife Division (WD) (2000). *Wildlife Division Policy for Collaborative Community Based Wildlife Management.* Technical Report, Accra, Ghana.

[14] Wittemyer, G., Elsen, P., Bean, W., T., Coleman, A., Burton, A. and Brashares, J., S. (2008). Accelerated human population growth at protected area edges. *Science*. 321, 123–126.

[15] Sergio, F., Caro, T., Brown, D., Clucas, B., Hunter, J., Ketchum, J., Mchugh, K. and Hiraldo, F. (2008). Top predators as conservation tools: ecological rationale, assumptions, and efficacy. *Annu. Rev. Ecol. Evol. Syst*. 39, 1–19.

[16] Woodroffe, R. and Ginsberg, J., R. (1998). Edge effects and the extinction of populations inside protected areas. *Science*. 280, 2126–2128.

In: Sustainable Utilization
Editors: E. D. Wiafe and F. S. Arku
ISBN: 978-1-63482-896-3

Chapter 2

DENSITY AND DISTRIBUTION OF BIRDS IN THE XAVI ECOTOURISM SITE, GHANA

Emmanuel Danquah* and Evans Atsu Abormega
Department of Wildlife and Range Management,
Faculty of Renewable Natural Resources,
Kwame Nkrumah University of Science and Technology,
PMB, Kumasi, Ghana

ABSTRACT

The study provides habitat information on density and distribution of bird species in Xavi Ecotourism Site (XES) in the Akatsi district of the Volta Region, Ghana. This may be used as a basis for improved tourism at XES and for future monitoring of climate change impacts on the avian community. Using the Timed Species-Count Method, one hundred and one bird species were identified. Distance to the Lotor River greatly influenced the density and diversity of birds. Major human activities identified in the XES including fishing, hunting, vegetable farming, honey gathering with fire and charcoal production may impact negatively on the birds.

Keywords: Xavi, ecotourism, river, vegetation, birds, distribution

---

* Corresponding Author: emmanueldanquah@yahoo.com, emmanueldanquah@gmail.com; Phone: +233-244 742 385.

1. INTRODUCTION

The avifauna in the Xavi Ecotourism Site (XES) at Xavi in the Akatsi District, Volta Region is believed to constitute over 100 different species of birds. Although local people have taken initiatives to conserve and manage the general biodiversity of the area and also generate revenue through ecotourism and local initiatives, not much has been done to document the bird species or monitor trends in density and distribution, which is important for evaluating biodiversity dynamics in future climate change impact studies. [1] Also, tour guides are unable to accurately identify most bird species or their habitats; hence, tourists hardly enjoy tour shifts or services rendered to them.

Information on distribution of birds can be used by management to enhance their revenue generating capacity through informed tourism, where tourists maximize bird watching experience by visiting only key bird sites. The XES can also take appropriate conservation decisions based on the distribution of key bird habitats and the factors that impact on them. According to [1], ecology of birds (including factors that affect their habitat and distribution) yield important data for bird conservation.

The study aims at providing habitat information on densities and distribution of bird species that may be used as a basis for improved tourism at XES and for future monitoring of climate change impacts on the avian community. Specifically, the objectives of the study are to provide baseline information on diversity and density of birds in the Xavi Eco-tourism Site and assess the impact of the Lotor River on density and distribution of birds. The Lotor River is the main water body and tourist attraction at XES and its influence on bird species has important consequences for promoting ecotourism and biodiversity conservation in the area, especially in the era of climate change.

2. STUDY AREA

The Xavi Eco-tourism Site (XES) is about 32 km^2 and it is located in the Akatsi District of the Volta Region in Ghana, West Africa. The Lotor River is the main water body in the XES and is a tributary of River Waya. The widest part of the river measures about 30metres and traverses the entire length of the XES over a distance of about 9km (Figure 1). Many tourists delight in the opportunity to paddle in canoes along the banks.

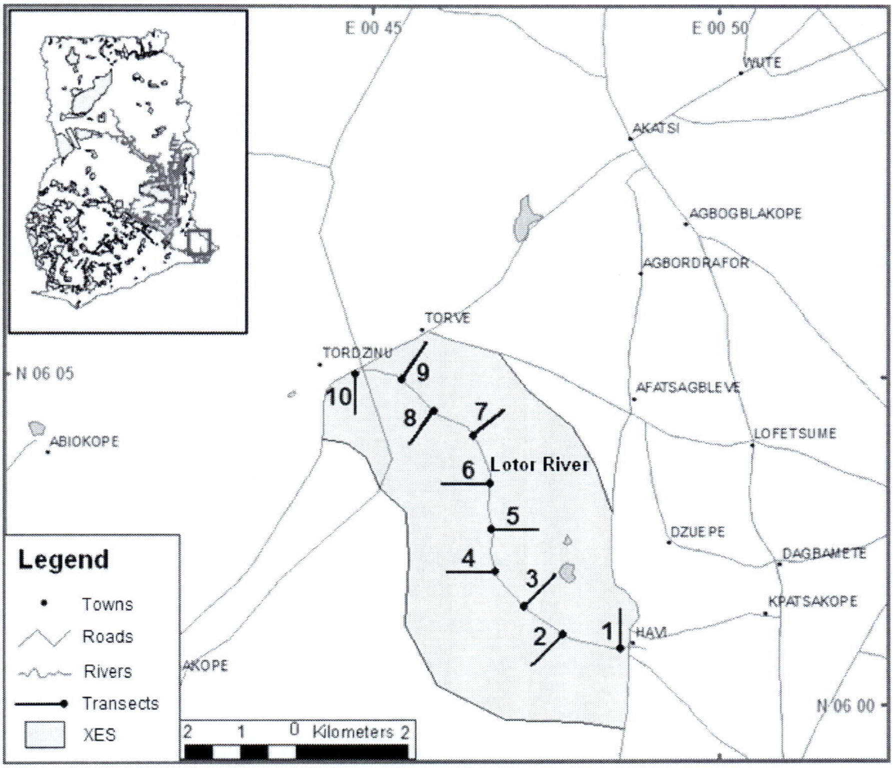

Figure 1. Map of Xavi Ecotourism Site (XES) showing the distribution of transects. The inset map shows the location of the site in south eastern Ghana.

The inhabitants of Xavi are primarily farmers and fishermen and rely on the Lotor River for fish as well as the arable lands in the catchment for growing vegetables. The river provides water for irrigating the farms, especially in the dry season. Major off season commercial activities include charcoal burning and wild honey hunting.

The XES has three main vegetation types: riverine forest, Guinea savanna woodland and grassland. The riverine forest occurs on the immediate banks of the Lotor River and harbours different species of birds, monkeys, reptiles and other biodiversity. It forms high-crowned riparian forest fragments with closed canopy at some sections with loose undergrowth. Plant species common to the riverine forest are red and white mangrove and *Mitragina* species. The Guinea savanna woodland and grassland occur further from the Lotor River and form the main vegetation type of the XES. This zone is dominated by plant species typical to the savanna including *Afzelia Africana, Pakia clappertoniana,*

Daniellia olivera, and *Vitellaria paradoxa.* Human interference usually takes place here and some areas are disturbed through farming activity, fishing camps and repeated bushfires.

3. Methods

3.1. Distribution of Transects

The bird survey was done with a network of 10 transects, each 1 km in length. Based on Geographic Information System (GIS) applications, the centrally flowing Lotor River was used as a baseline and systematically delineated into 1-km intervals. The intersections of the intervals along the river formed the starting point for each transect. The Lotor River is about 9km long and this gave 10 transects conforming to a *systematic segmented line transect* design (Figure 1). Transect orientation was perpendicular to the Lotor River.

3.2. Data Collection and Analysis

The Timed Species-Count (TSC) method was employed to count birds in the XES. This is a combination of direct and indirect survey techniques which helps to identify species and assign index of relative abundance to them. [2]

Two sessions of surveys were conducted, and four sets of data on birds were collected simultaneously. TSC was conducted at dawn, noon and late evening on all the ten transects individually. As a result, 30 surveys were done during the first session for 10 days. A second (repeat) session commenced three days after the end of the first session. Thus, a total of 60 counts were conducted at the end of the survey.

Transect data on birds was categorized into 250m interval gradients from the Lotor River resulting in four bands. The goal was to build a perpendicular gradient from the Lotor River along which the effect of the river can be assessed in relation to bird density and species diversity.

Regression analyses and the Friedman Test were conducted to check for relationships and differences respectively in bird density and species diversity along the perpendicular gradient from the Lotor River. The software EstimateSWin800 Version 7.5.0 [3] was used to estimate indices of bird diversity.

4. RESULTS

4.1. Diversity and Density of Bird Species

Nine hundred and thirteen bird signs, representing 101 species(Appendix 1) of the eight Guilds; Insectivore (31 species), Graminivore (20), Piscivore-Insectivore (16), Frugivore (10), Frugivore-Insectivore (9), Nectarivore-Insectivore (7), Omnivore (5) and Raptors (3) were confirmed in the Xavi Ecotourism Site during the survey.

The most commonly catalogued signs were those of the common bulbul (Pycnonotusbarbatus), which represented a density of 6.0 bird signs per km, the vinaceous dove (Streptopeliavinacea; 4.9), the bronze manikin (Lonchuracucullata; 4.7) and the village weaver (Streptopeliavinacea; 4.6). Rarer birds, including the zittingcisticola (Cisticolajuncidis), reef egret (Egrettagularis), rose-ringed parakeet (Psittaculakrameri), red-tailed greenbul (Crinigercalurus), glossy starling (Lamprotornispurpureus), grey-backed camaroptera (Camaroptera brachyuran), greater honey guide (Indicator indicator), black-crowned sparrow hawk (Eremopterixnigriceps) had densities of less than 0.5 bird signs per km (Appendix 1).

Bird density (Friedman Test: Chi-sq ($\chi 2$) = 3.231, d.f. = 3, $P>0.05$) and species diversity (Friedman Test: Chi-sq ($\chi 2$) = 3.346, d.f. = 3, $P>0.05$) were significantly different across the defined gradients from the Lotor River. Density of birds increased with distance from the Lotor River whilst species diversity decreased along the same gradient (Table 1).

Table 1. Bird density and species diversity within 250m interval gradients from the Lotor River

Gradient (meters)	Bird Density (per km)	Species Diversity (Simpsons mean)
0 – 250	18.10	77.57
250 – 500	21.90	83.47
500 – 750	25.20	56.17
750 – 1000	26.10	46.37

4.2. Impact of the Lotor River on Bird Density and Diversity

Distance to the Lotor River significantly influenced density (r^2=0.792, $P<0.01$) and species diversity (r^2=-0.796, $P<0.01$) of birds. The regression models showed increasingbird density (Figure 2) but decreasingspecies diversity (Figure 3) with increasingperpendicular distance from the Lotor River. There seems to be a threshold distance of about 200 meters from the Lotor River, beyond which most bird species are rapidly lost with increasing distance (Figure 3).

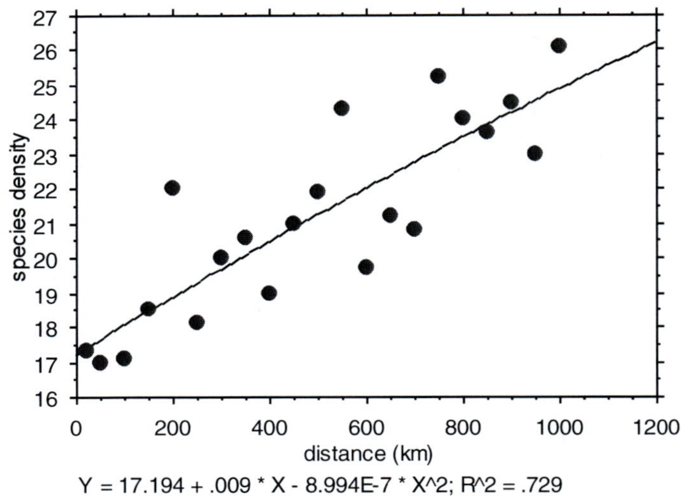

Figure 2. Relationship between distance from the Lotor River and density of birds.

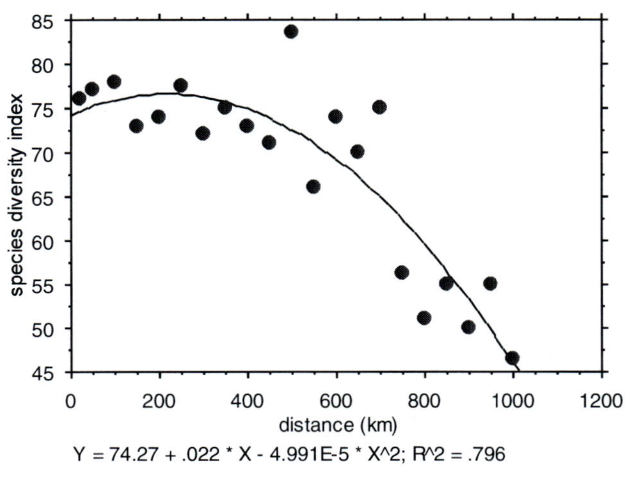

Figure 3. Relationship between distance from the Lotor River and diversity of birds.

Major human activities identified in the XES consist of fishing, hunting, vegetable farming, honey gathering with fire and charcoal production. With the exception of some fishing activity recorded along the banks of the Lotor River, most of these human interferences including bushfires were encountered in the savannah woodlands further from the river.

5. Discussion

The considerable density and diversity of birds within the XES is encouraging and illustrates the potential value of off-reserve forests for conservation. This observation strengthens the rationale for including community-managed forest approaches in both local and regional conservation efforts.

The study suggests that riverine forests/vegetation that buffer the Lotor River contain significant bird diversity and may serve as important sites for biodiversity conservation. There were important differences in the density and diversity of birds as one moved away from the Lotor River into the savannah woodland and grassland. The observed pattern of increasing density but decreasing diversity from the Lotor River suggest a general gradient of diminishing vegetation diversity from the riverine forest surrounding the Lotor River to the less diverse woodland and grassland that occur further (> 200 metres) from the river. Riverine vegetation and stream corridors normally hold unusual promise for biological conservation because diversity of plants and animal species is often disproportionately high. [4] The results may present additional evidence that vegetation changes potentially have important consequences for concomitant fauna assemblages [5], with significant differences being apparent between forest and non-forest landscapes. [6-10]

Alternatively, there are a number of relatively large communities: Azametikope, Akogokorpe, Kudzorkope, Hiatorkorpe, dotted along the periphery of the XES. These communities present a source of relatively high human disturbance along the boundary, including farming, bush burning by palm wine tappers and honey hunters and also cutting down of trees by charcoal burners that could impact negatively on the birds. Villagers also enter the XES to conduct communal hunting of small rodents and search for other wild resources including fuel wood. Compared to the riverine forests, most of these human interferences were encountered in the savannah woodlands further from the Lotor River and could contribute to the differences in species diversity and density of birds in the different vegetation types. Other suspected factors could of course be gradients in the distribution of food, nesting site or even competition.

A number of species react to variations in the structure of vegetation similarly to human pressure, by underutilizing non-forest landscapes or areas dominated by humans [11-13], but the strength of this reaction differs for different species. [12] Many wildlife species shift habitat under long-term, intense human activity which result in reduced access to resources. [14] Habitat-sensitive species, migratory species and dietary specialists may face major resource adaptation challenges, dispersal and predator avoidance within human dominated landscapes resulting in decline in species diversity or local extinctions. [15]Although forest conversion to non-forest vegetation types may not support the conservation of diverse bird species, few prolific and highly resilient generalist as well as several opportunistic species may adapt, and even thrive, within less diverse environments [16, 17] or human dominated landscapes.

Conclusion and Recommendations

At XES, it is likely that a diverse community of specialist species occur in low densities in the riverine forests bordering the Lotor River whilst a few generalist species occur in high densities further from the river in the woodland and grassland vegetation. However, as the ecology of African forest birds is still poorly known in many areas, only speculations can be made. Given the level of human activities that occur in the XES, assessing the land use dynamics with wildlife presence in general is critically important for effective conservation planning [18], especially in the current era of global climate change. Efforts to conserve bird diversity must therefore involve the close collaboration of local communities to promote resilient and sustainable land use systems. This will facilitate the conservation of biodiversity while enabling local communities to meet their livelihood needs and continue their traditional practices. [19]

The Lotor River could offer something that is perhaps unique to XES. There are many tourists who would delight in the opportunity to paddle and spot birds along the banks. Camp-sites could be established along the river banks. Tour operators could restrict the number of tourist entering that stretch of the banks at any one time to avoid over-crowding and degradation of key riverine habitat. If this tourist experience is well managed, it could form the central feature on which to base publicity for XES.

ACKNOWLEDGMENT

We thank the chief and people of Xavi, including our tour guides; Christian and Hiator for their assistance. Mr. Robert Nutakor of Ashanti African Tours, assisted with various aspects of this research including field work.

APPENDIX 1. DENSITY OF BIRD SPECIES IN THE XAVI ECOTOURISM SITE

Common Name	Scientific Name	Total Signs	Density (per km)	Rank
African Green Pigeon	*Treron calvus*	31	3.1	17
African Grey Hornbill	*Tockus nasutus*	32	3.2	15
African Jacana	*Actophilornis africana*	34	3.4	13
African Moustached Warbler	*Melocichiamentalis*	8	0.8	73
African Paradise Flycatcher	*Terpsiphoneviridis*	6	0.6	81
African Pied Hornbill	*Tockusfasciatus*	5	0.5	86
African Pied Wagtail	*Motaciliaaguimp*	21	2.1	37
African Thrush	*Turduspelios*	17	1.7	48
Bar Breasted Firefinch	*Lagonostictarufopicta*	15	1.5	53
Black Crowned Sparrow Lark	*Eremopterixnigriceps*	4	0.4	94
Black Kite	*Milvusmigrans*	40	4	8
Black-Billed Wood Dove	*Turturabyssinicus*	19	1.9	40
Blackcap Babbler	*Turdoidesreinwardti*	7	0.7	76
Black-crowned Thagra	*Tehagrasenegala*	10	1	66
Black-Winged Bishop	*Euplecteshordeaceus*	28	2.8	22
Blue bellied Roller	*Coraciacyanogaster*	10	1	66
Blue breasted Kingfisher	*Halcyon malimbica*	7	0.7	76

Appendix 1. (Continued)

Common Name	Scientific Name	Total Signs	Density (per km)	Rank
Blue Quail	*Coturnixadansonii*	8	0.8	73
Blue-billed Malimbe	*Malimbusnitens*	5	0.5	86
Blue-Headed Wood Dove	*Turturbrehmeri*	28	2.8	22
Broadbilled Roller	*Eurystomusglaucurus*	13	1.3	57
Bronze Manikin	*Lonchuracucullata*	47	4.7	3
Brown Babbler	*Turdoidesplebejus*	28	2.8	22
Capuchin Babler	*Phyllanthusatripennis*	22	2.2	36
Cattle Egret	*Bubulcus ibis*	17	1.7	48
Collared Sunbird	*Anthreptescollaris*	27	2.7	26
Common Bulbul	*Pycnonotusbarbatus*	60	6	1
Common Swift	*Apus apus*	36	3.6	12
Common Wattle- Eye	*Platysteiracyanea*	6	0.6	81
Copper Sunbird	*Nectariniacuprea*	29	2.9	20
Didric Cuckoo	*Chrysococcyxcaprius*	19	1.9	40
Double Spurred Francolin	*Francolinusbicalcaratus*	45	4.5	5
Double-Toothed Barbet	*Lybiusbidentatus*	18	1.8	44
Goliath Heron	*Ardea goliath*	5	0.5	86
Greater Honeyguide	*Indicator indicator*	2	0.2	99
Green Wood-Hoopoe	*Phoeniculuspurpureus*	12	1.2	60
Green-Backed Heron	*Butoridesstriata*	12	1.2	60
Grey Backed Camaroptera	*Camaroptera brachyuran*	4	0.4	94
Grey Headed Kingfisher	*Halcyon leucocephala*	11	1.1	63
Grey Woodpecker	*Dendropicosgoertae*	27	2.7	26
Heuglins Masked Weaver	*Ploceusheuglini*	9	0.9	72
Klaas Cuckoo	*Crysococcyxklaas*	18	1.8	44
Laughing Dove	*Streptopeliasenegalensis*	30	3	19
Leaf Love	*Pyrrhurusscandens*	31	3.1	17
Lesser Moorhen	*Gallinulaangulata*	5	0.5	86
Levaillant's Cuckoo	*Clamatorlevaillantii*	23	2.3	30
Little Bee-Eater	*Meropspusillus*	17	1.7	48
Little Bittern	*Ixobrychus minutes*	7	0.7	76

Common Name	Scientific Name	Total Signs	Density (per km)	Rank
Little Blackcap Tchagra	*Tchagraminuta*	21	2.1	37
Little Egret	*Egrettagarzetta*	16	1.6	51
Lizard Buzzard	*Kaupifalcomonogramicus*	28	2.8	22
Malachite Kingfisher	*Alcedocristata*	12	1.2	60
Melodious Warbler	*Hippolaispolyglotta*	24	2.4	29
Northern Grey-Headed Sparrow	*Passer griseus*	44	4.4	7
Northern Red Bishop	*Euplectesfranciscacus*	15	1.5	53
Olive Sunbird	*Cyanomitraolivaceus*	21	2.1	37
Olive-bellied Sunbird	*Nectariniachloropygia*	5	0.5	86
Oriole Warbler	*Hypergerusatriceps*	13	1.3	57
Pale Flycatcher	*Bradornispallidus*	10	1	66
Piapiac	*Ptilostomusafer*	18	1.8	44
Pied Crow	*Corvusalbus*	27	2.7	26
Pied Kingfisher	*Cerylrudis*	38	3.8	10
Pin Tailed Whydah	*Viduamacroura*	23	2.3	30
Piping Hornbill	*Ceratogymnafistulator*	7	0.7	76
Plain-Backed Pipit	*Anthusleucophrys*	11	1.1	63
Purple Glossy Starling	*Lamprotornispurpureus*	2	0.2	99
Purple Heron	*Ardeapurpurea*	15	1.5	53
Pygmy Kingfisher	*Ceyxpicta*	23	2.3	30
Pygmy Sunbird	*Anthreptesplaturus*	23	2.3	30
Red headed Lovebird	*Agapornispullarius*	7	0.7	76
Red Tailed Greenbul	*Crinigercalurus*	4	0.4	94
Red-Billed Firefinch	*Lagonostictasenegala*	45	4.5	5
Red-eyed Dove	*Streptopeliasemitorquata*	34	3.4	13
Rose-Ringed Parakeet	*Psittaculakrameri*	3	0.3	98
Scarlet-Chested Sunbird	*Chalcomitrasenegalensis*	5	0.5	86
Senegal Coucal	*Centropussenegalensis*	38	3.8	10
Senegal Parrot	*Poicephalussenegalus*	19	1.9	40
Shikra	*Accipiter bardius*	32	3.2	15
Simple Leaflove	*Chlorocichla simplex*	10	1	66
Slender-billed Weaver	*Ploceusluteolus*	10	1	66
Splendid Sunbird	*Nectariniacoccinigaster*	16	1.6	51
Squacco Heron	*Ardeolaralloides*	5	0.5	86
Swallow-tailed Bee Eater	*Meropshirundineus*	6	0.6	81

Appendix 1. (Continued)

Common Name	Scientific Name	Total Signs	Density (per km)	Rank
Tawny Flanked Prinia	*Priniasubflava*	8	0.8	73
Tropical Boubou	*Laniariusaethiopicus*	14	1.4	56
Village Weaver	*Ploceuscucullatus*	46	4.6	4
Vinaceous Dove	*Streptopeliavinacea*	49	4.9	2
Vitelline Masked Weaver	*Ploceusvelatus*	11	1.1	63
Western Grey Plantain Eater	*Criniferpiscator*	39	3.9	9
Western Reef Egret	*Egrettagularis*	4	0.4	94
White Rumped Swift	*Apus caffer*	23	2.3	30
White-faced whistling duck	*Dendrocygnaviduata*	19	1.9	40
White-throated Bee Eater	*Meropsalbicollis*	23	2.3	30
Woodland Kingfisher	*Halcyon senegalensis*	10	1	66
Yellow Crowned Bishop	*Euplectesafer*	18	1.8	44
Yellow Fronted Tinkerbird	*Pogoniuluschrysoconus*	29	2.9	20
Yellow-Billed Shrike	*Corvinella corvine*	13	1.3	57
Yellow-Fronted Canary	*Serinusmozambicus*	5	0.5	86
Yellow-Mantled Widowbird	*Euplectesmacrourus*	6	0.6	0.6
Yellow-RumpedTinkerbird	*Pogoniulusbilineatus*	6	0.6	81
ZittingCisticola	*Cisticolajuncidis*	3	0.3	98

REFERENCES

[1] Bibby, C., Jones, M. and Marsden, S. (1998) Expedition Field Techniques: Bird Surveys. Expedition Advisory Centre, London, UK.

[2] Pomeroy, D., E. and Tengecho, B. (1986). Studies of birds in a semi-arid area of Kenya. III – the use of 'timed species-counts' for studying regional avifaunas. *Journal of Tropical Ecology*, 2: 231-247.

[3] Colwell, R. K. 2005. EstimateS: statistical estimation of species richness and shared species from samples. Version 7.5. User's guide and application. University of Connecticut, Storrs. Available from http://purl.oclc.org/estimates (accessed November 2006).

[4] Noss, R., F. (1983). A regional landscape approach to maintain diversity. Bioscience, 33: 700-706.

[5] Cullen, L., Bodmer, E. R. and Valladares-Padua, C. (2001). Ecological consequences of hunting in Atlantic forest patches, São Paulo, Brazil. Oryx, 35:137-144.

[6] Schelhas J. and Greenberg R. 1996. Forest Patches in Tropical Landscapes. Island Press, Washington, DC, pp. 426.

[7] Laurance, W. and Bierregaard, R., O. (1997). Tropical Forest Remnants: Ecology, Management and Conservation of Fragmented Communities. University of Chicago Press, pp. 616.

[8] Daily, G., C., Ceballos, G., Pacheco, J., Suz´an, G. and S´anchez-Azofeifa, A. (2003). Countryside biogeography of Neotropical mammals: conservation opportunities in agricultural landscapes of Costa Rica. *Conservation Biology17*:1814–1826.

[9] Matlock, R., B., Jr., Rogers, D., Edwards, P., J. and Martin, S., G. (2002). Avian communities in forests fragments and reforestation areas associated with banana plantations in Costa Rica. A*griculture Ecosystem Environment*. 91: 199–215.

[10] Harvey, C., A., Gonzalez, J. and Somarriba, E. (2006). Dung beetle and terrestrial mammal diversity in forests, indigenous agroforestry systems and plantain monocultures in Talamanca, Costa Rica. *Biodiversity and Conservation* 15:555–585.

[11] Sutherland, W., J. and Crockford, N., J. (1993). Factors affecting the feeding distribution of Red-breasted Geese Brantaruficolliswintering in Romania. *Biological Conservation* 63: 61–65.

[12] Gill, J., A., Sutherland, W., J. and Watkinson, A., R. (1996). A method to quantify the effects of human disturbance on animal populations. *Journal of Applied Ecology* 33: 786–792.

[13] Beale, C., M. and Monaghan, P. (2004). Human disturbance: people as predation-free predators? *Journal of Applied Ecology* 41: 335–343.

[14] Frid, A., and L. Dill. 2002. Human-caused disturbance stimuli as a form of predation risk. Conservation Ecology 6: http://www. consecol.org /vol6/iss1/art11.

[15] Tait, L., Lester, H. and Birchwood, M. (2005) Design of the Birminghamearly detection in untreated psychosis trial (REDIRECT):

cluster randomized controlled trial of general practitioner education in detection of first episode psychosis. *BMC Health Services Research* 5: 1-8.

[16] Barnes, R., F., W., Azika, S. and Asamoah-Boateng, B. (1995). Timber, cocoa, and crop-raiding elephants: A preliminary study from southern Ghana. *Pachyderm* 19: 33-38.

[17] Fonseca, G., A., B. and Robinson, J., G. (1990). Forest size and structure: competitive and predatory effects on small mammal communities. *Biological Conservation*. 53: 265-294.

[18] Sutherland, W., J. (1998). The importance of behavioural studies in conservation biology. Animal Behaviour56: 801–809.

[19] Colchester, M. (2004). Conservation policy and indigenous peoples. *Environmental Science Policy* 7: 145–153.

In: Sustainable Utilization
Editors: E. D. Wiafe and F. S. Arku
ISBN: 978-1-63482-896-3

Chapter 3

BOTANICAL ASSESSMENT IN A COMMUNITY FOREST RESOURCE MANAGEMENT AREA, GHANA

Edward D. Wiafe*
Department of Environmental and Natural Resources,
Presbyterian University College, Ghana

ABSTRACT

This chapter involves a botanical assessment of Bodaa Community Resource Management Area in the Brong-Ahafo Region of Ghana. One vegetation sample plot (100m by 100m) was associated with and placed systematically along transects in the various vegetation types. Thus, 30 quadrats were laid. Sample plots were demarcated with the help of ranging poles and distances measured with a linear tape. The enumeration team was made up of a tree spotter, a recorder and two assistants. Moving clock-wisely, all plants including trees with girth of 31.5 cm (dbh>/=10cm) were systematically identified, measured and recorded. Interactions between residents of target communities and the biotic elements of nature around them (including indigenous knowledge systems, practices and technologies relating to biological elements and processes) were investigated using participatory rural appraisal methods.

* Corresponding author: Edward D. Wiafe. Department of Environmental and Natural Resources, Presbyterian University College, Ghana, P. O. Box 393, Akropong-Akuapem, Ghana. E-mail: edward.wiafe@presbyuniversity.edu.gh.

This method employs direct administration of structured questionnaires and informal discussions with knowledgeable individuals in target communities. Several factors such as farming and illegal wood harvesting were identified as issues affecting tree distribution and conservation in the area. Recommendations have been made to facilitate community participation in the tree conservation in the area.

1. Introduction

The concept of communities managing natural resources is not new in Ghana. The Wildlife Division of the Forestry Commission has (since 1991) been developing and piloting a community based natural resource management program that has the potential to dramatically change the status of wildlife and other resources outside Protected Areas.

One of the broad approaches to reducing conflict of interest in the wildlife use in the protected area is by accepting that the communities are the moral owners of wildlife on their lands. This includes the liberalization and decentralization of conservation legislation, which together with institutional adjustments must restore incentives for rural to support and participate in the legal sustainable use of wildlife.

Harvesting of any natural resource products from the forest is expected to have an impact on proportion to the intensity of the off take [1] despite the numerous benefits by humans [2]. Local extinctions of vertebrates (whether browsers or top predators) are likely to eventually affect plant community through changes in inter-specific relations such as seed dispersal, predation and browsing. Changes in plants community will in turn have effects on other faunal components. This is because plants are the foundation of the food chain in all ecosystems. Plants form the actual physical substrate in which most organisms live- even those organisms that do not require the actual plant substance require the physical structure of a forest to survive, example for nest and roosting sites. Moreover, plants affect the actual environment; example the chemistry of fallen litter dictates the nutrients and chemical environment in which other organisms depend [3]. Plant roots also prevent erosion. Many plants have specific environmental requirement for survival and changes in soil chemistry, rainfall, temperature or humidity can affect the survival of various species differently which in turn can be used as good indicator of environmental changes [4].

The 1994 Forest and Wildlife policy enshrines the principle of conservation through sustainable development and recognizes the need to associate local communities with Protected Area management through the generation of benefits (e.g. natural resource utilization).

This approach is the basis of the philosophy of the Community Resource Management Areas (CREMA) program, promoting rural development predominantly through conservation management of wildlife.

This is with a rationale that if a protected area increasingly becomes an ecological island, the pressure from neighbouring communities to utilize the otherwise scarce natural resources contained within it will eventually overcome the ability to protect these resources. This could be achieved through integrating the use of natural resources into the existing land use practices. The program aims to:

1) support fringe communities to restore and manage habitat for the return of wildlife in the off-reserve areas through any appropriate mechanism
2) increase fringe communities' income from natural resources as an incentive for engaging in sustainable forest resource management practices
3) reduce pressure from communities on the protected areas for their livelihood needs
4) build the capacity of communities and the District Assembly to manage their forest resources.

As the wildlife situation begins to improve in off-reserve areas, the communities will be harvesting these resources in a sustainable manner. For this to be effective there is a need to determine what species are available in the areas and in which quantities; and determine the growth of the resources through time by periodic inventory through training of members on participatory inventory, bio-monitoring and evaluation.

The main goal is to provide floral information to support most CREMA programs such as restoration programs, formulation of byelaws, baseline information for monitoring, etc. The specific objectives of this floral survey were to:

- Assess the tree species composition and diversity occurring inside the community resource management area.

- Assess the type of use by the community members of the tree resources.
- evaluate threats to tree component in the area

2. THE CONCEPT OF COMMUNITY BASED NATURAL RESOURCE MANAGEMENT AREA

The Community Resource Management Area (CREMA) concept recognizes that all landscapes, those largely natural in form as well as those heavily modified by humans, are mosaics of different types of habitats which form a critical part of the range of most wildlife. The concept is therefore based on the establishment of areas where wildlife management is incorporated into existing land use. It is generally recognized that the current legislation does not provide sufficient incentive for farmers to care for and protect wildlife on their land. The result is that wildlife, for most farmers, is considered an open resource or in some instances a pest that in many cases is directly competing with their agricultural activities.

A case in point is the adjoining communities of Bodaa, Koraase and Biama in Jaman South District of the Brong Ahafo Region where for some time crop-raiding by marauding buffaloes was generating local animosity until the Wildlife Division was called in to sensitize the population on ways of reducing human-wildlife conflict and by so doing increased the social tolerance of the people towards the buffaloes and ultimately enhanced their conservation in the area by the establishment of a local CREMA.

According to [5] one of the most effective ways to obtain broad support for Community Based Natural Resource Management (CBNRM) programs is to integrate their planning and management with other program that deliver benefits in sustainable land management (such as protection of water resources or sustainable use of natural products). The CREMAs therefore aim at involving local communities and devolving wildlife management and utilization outside protected areas. The structures are designed so as to enable the promotion of wildlife as a viable complement to agricultural practices that are carried out in Ghana. It is meant to create a financial incentive for farmers to use and sustainably manage natural resources by devolving management rights and responsibilities to them. It uses an organizational structure that is based on existing decision-making structures and is consistent with local land tenure relationships.

To advance the process of establishing such a CREMA and sustainably manage the wildlife on communal lands of the three communities, a baseline floral survey and participatory floral assessment training was conducted for the key executive members of Bodaa CREMA. This report presents the findings of the survey.

2.1. Study Area

The study took place in the Brong-Ahafo Region of Ghana between longitudes 7°440'N-7°510'N and 2°450'W-2°510'W.

The major occupation in the area is farming on main food crops such as maize, cassava, yam and on cash crops such as cashew, palm etc.

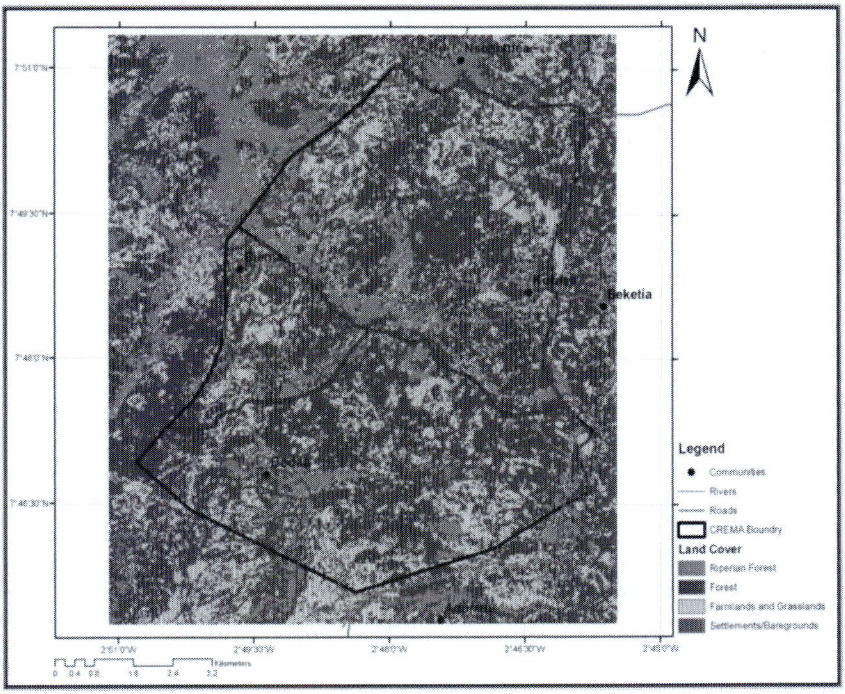

Figure 1. Landcover map of the Bodaa CREMA.

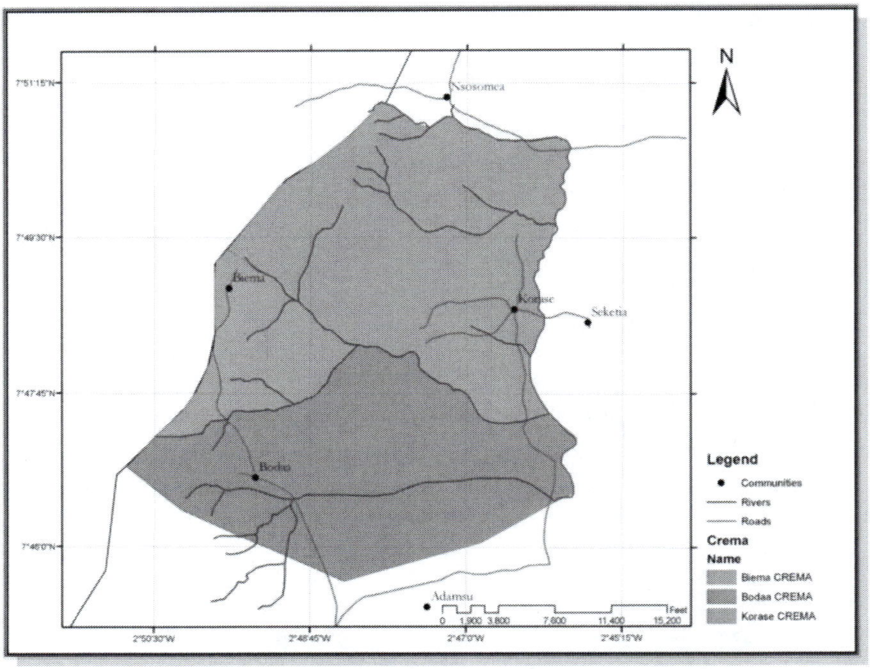

Figure 2. Map of Bodaa CREMA showing different management units.

2.1.1. Plot Demarcation and Distribution of Quadrats (Plots)

As a basis for comparison, one vegetation sample plot (100m by 100m) was associated with and placed systematically along transects in the various vegetation types. Thus, 30 quadrats were laid. Sample plots were demarcated with the help of ranging poles and distances measured with a linear tape. An access line was cut through the middle of the plot to facilitate movement through the plot, as shown in Figure 3.

2.1.2. Enumeration Procedure

Habitat assessment and vegetation inventories in respect of wildlife welfare and growth within and around the CREMA comprised both qualitative and quantitative methods [6].

All habitat features including vegetation and land-use types were included in the survey to make a complete inventory for the CREMA. The enumeration team was made up of a tree spotter, a recorder and two assistants.

Moving clock-wisely, all plants including trees with girth of 31.5 cm (dbh>/=10cm) were systematically identified, measured and recorded.

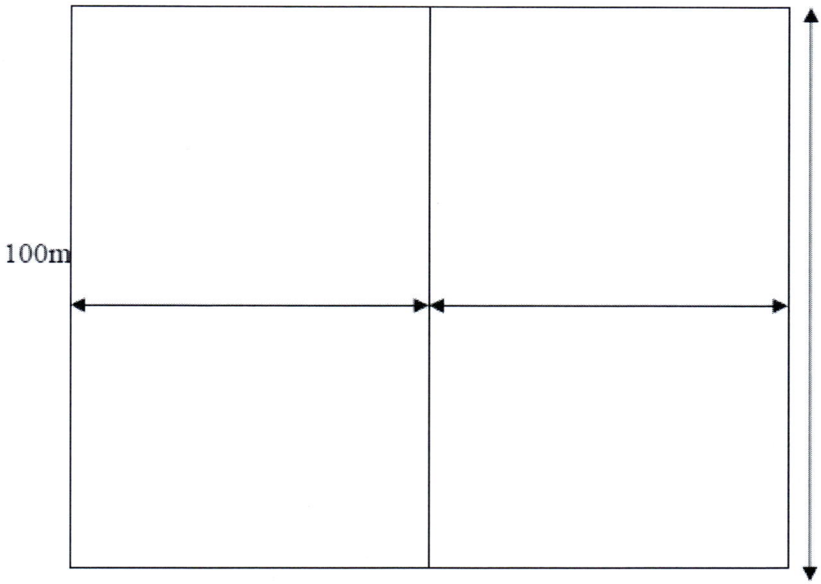

Figure 3. Shape and dimensions of a sample plot, showing the boundary distances and the access line at the middle of the plot. It covers an area of 1 hectare ($10{,}000m^2$).

All trees encountered were identified to the species level and specimens of unidentified trees collected and sent to Resource Management Support Centre's herbarium in Kumasi for identification.

Species local names were called out to the recorder who in turn called back the same information to ensure that the right information was recorded.

2.2. Ethno-Biological and Socioeconomic Assessment of the CREMA

Effective conservation requires data on both biodiversity and resource use. The CREMA offer opportunity to sustainably manage community resources in a manner that perpetuate fauna, flora and biophysical resources within community landscapes.

Understanding community's use of resources, both past and present, is crucial for determining future uses, best practices and opportunities for livelihoods diversification and environmental health improvement.

Interactions between residents of target communities and the biotic elements of nature around them, including indigenous knowledge systems, practices and technologies relating to biological elements and processes were investigated using participatory rural appraisal methods, employing direct administration of structured questionnaires and informal discussions with knowledgeable individuals in target communities.

Specifically, the assessment covered the following areas:

- identification species (both flora and fauna) which are most favoured for food, medicine, craft and other purposes in time.
- varieties, types and parts of flora and fauna species utilised by community residents over time

2.3. Training of CREMA Members in Participatory Inventory, Bio-Monitoring and Evaluation

The survey team for the CREMA was kept permanent to ensure consistencies; hence, a permanent group of community members appointed by the communities were pre-trained to participate in the survey.

These trainers are expected to train more community members to do their own inventory, bio-monitoring and evaluation in the future.

Potential candidates were interested in conservation activities and chosen from existing volunteer groups assisting in conservation activities in the community. Technically skilled assistants were recruited to improve data collection.

3. RESULTS

3.1. Tree Density and Diversity

In Bodaa, Korase and Biama management units i.e., Community Resource Management Committees (CREMC), trees enumerated were 45, 46 and 19 per hectare respectively.

The number of species enumerated in Bodaa was 43, Korase 47 and Biama, 44. Table 1 indicates the details of diversity and evenness of tree enumerated in the proposed CREMA.

3.2. Distribution of Number of Trees in Diameter Classes of the Enumerated Trees in the Proposed Bodaa CREMA

The study revealed that the greater number of trees enumerated in the proposed CREMA fall within the smaller diameter classes i.e. between 10 to 30 cm diameter classes. In Bodaa and Korase CREMCs, tree numbers decreased with increasing diameter (Figure 3 and 4) but at Biama though a similar trend was observed relatively high number of trees was also recorded in the higher diameter class (60.5 and above) (Figure5).

Table 1. Indices of abundance of trees enumerated in the proposed Bodaa CREMA

CREMC	Number of Species	Number of individuals	Shannon Index	Simpson Index	Evenness
Bodaa	43	321	2.95	0.90	0.77
Korase	47	323	3.15	0.93	0.82
Biama	44	136	3.47	0.96	0.92

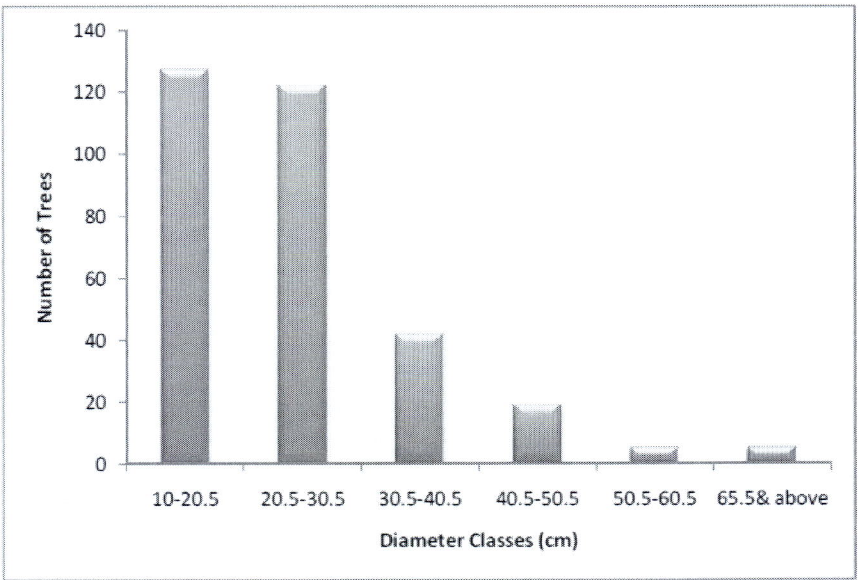

Figure 4. Diameter distribution of trees enumerated in the proposed Bodaa CREMA.

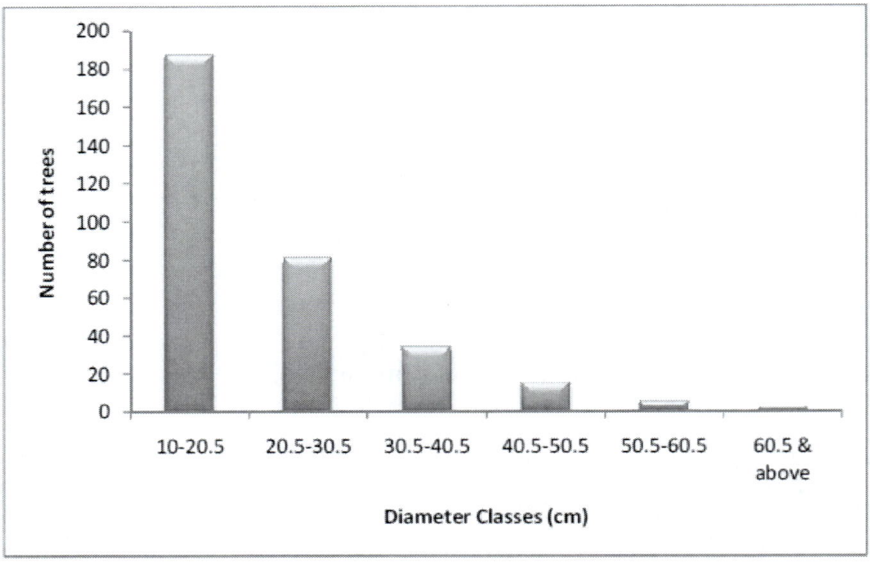

Figure 5. Diameter distribution of trees enumerated in the proposed Korase CREMA.

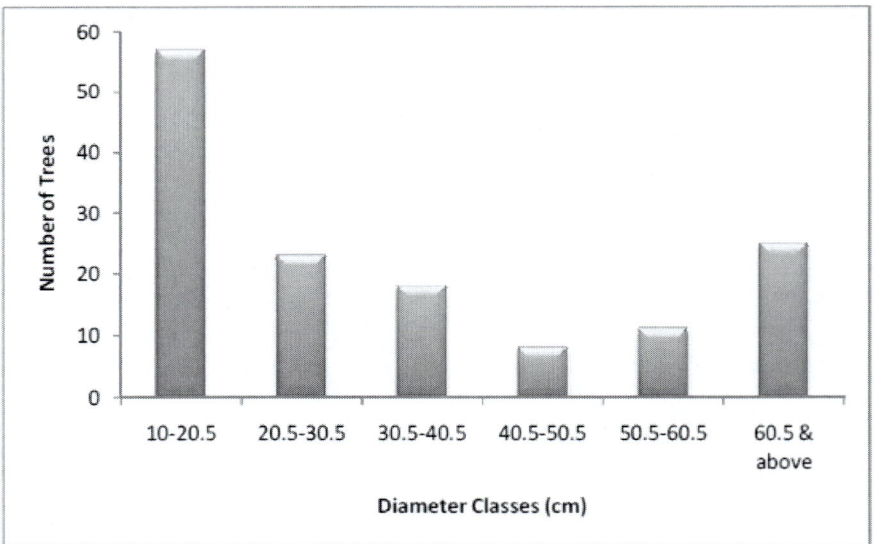

Figure 6. Diameter distribution of trees enumerated in the proposed Biama CREMA.

3.3. General Uses of Tree Resources in the Proposed Bodaa CREMA

The study found that wood is one of the oldest sources of energy and commonest service material known to the people of the Bodaa CREMA and has been used for several years.

Factors such as family size, cost of wood, season, type of cooking device, alternative energy sources and the type of wood determine the level of wood consumption. It was found that between 95-97% of the people depend on wood or charcoal as a source of domestic energy.

Some of the tree species were also used as a source of medicine and 99% of all men interviewed confirmed the use of tree resources as sources of traditional medicine. Some trees were said to be good for moisture retention and 70% of the farmers stated that those trees were good for farming.

Other uses of tree resources include the poles, fencing, sale of firewood for money and conversion of wood to charcoal. Table 2 shows the uses of 10 commonest trees and their uses by the people of the Bodaa CREMA.

3.4. Factors Affecting Tree Conservation in the Proposed Bodaa CREMA

The main human activities identified in the proposed Bodaa CREMA that affect tree resources include farming, firewood extraction, charcoal production and chainsaw lumber production. The land cover classification indicated four major land cover types: forest, farmland/grassland, settlement/bare-ground and swampy areas (Figure 1). Farming concentrated on the fertile forest areas of the CREMA, especially Biama where cocoa and other forest crops grow well. At Biama and Kroase, the major cash crop grown here is the savannah friendly crops, especially cashew-nut and other farm crops like yam, cassava etc.

Charcoal production and firewood extraction are very pronounced activities in the proposed CREMA. Several piles of firewood are the common scene in the area, where firewood was piled up for buyers to purchase to other areas.

Moreover, citizens convey fire wood for domestic consumption while others also convert it to charcoal for sale in the cities.

Conversion of tree logs to lumber with the use of chainsaw machine was the only source of lumber for house roofing and other joinery activities. Others also take advantage and send the lumber to the cities for sale (Figure 7).

Table 2. Some uses of trees enumerated in the Bodaa CREMA

Local Name	Scientific name	Uses	% Yes	% No
Nyamedua	*Altonia boonei*	Medicinal, Carving	100	0
Kane	*Anogeissus leiocarpus*	Firewood, poles	75	25
Osena	*Dialium gueneense*	Fruit, pestles	95	5
Konkroma	*Morinda lucida*	Medicinal, fire wood	80	20
Osrono	*Parkia bicolor*	Fruits, condiments	100	0
Osanya	*Daniellia oliveri*	Lumber, firewood, shade	60	40
Ongo	*Terminalia*	Medicinal, fire wood	77	33
	Glaucescens			
Dwindwera	*Lecaniodiscus*	Fire wood, medicinal	68	32
	Cupanioides			
Sasamasa	*Newbouldia laevis*	Medicinal, good for cocoa farm	100	0
Mahogany	*Khaya senegalensis*	Medicinal, wood	93	7

Figure 7. (Continued).

Figure 7. Chain saw activities observed in the proposed Bodaa CREMA.

3.5. Training of Community Member

In all 15 members of the community were trained on basic plant enumeration procedures. Five members were appointed by the community from each proposed CRMC and were trained on basic sampling procedures, plot demarcation, tree identification and measurements. The trained members were made to participate in the inventory of the botanical assessment. Trained members were evaluated at the end of the exercise to ascertain the level of knowledge acquisition, application and skills. Figure 8 is a picture showing the training procedures of the CREMA members.

3.6. Tree Species Composition

Table 3 contains the scientific name and the respective local names used in the proposed Bodaa CREMA. The nomenclature of scientific names was after [7] and [8].

Figure 8. Members of the communities that were trained on basic enumeration procedures from the various CRMCs.

Conclusion

In spite of the numerous uses of tree resources in the area, strong economic forces coerced the local people to over harvest the trees either for commercial purposes or domestic use.

Table 3. Composition of tree species encountered in the proposed Bodaa CREMA

Scientific name	Local name
Acacia spp.	Acacia
Afzelia Africana	Papao
Albizia adianthifolia	pampena
Albizia ferruginea	Awiemfosamina
Albizia zygia	Okro
Alstonia boonia	Nyamedua
Alstonia boonia	Sinuro
Anogeissus leiocarpus	Kane
Anthocleist anobilis	Bontodie
Antiaris toxicaria	Kyenkyen
Blighia sapida	Akye
Bombax bounopozense	Akonkodie
Bridelia micrantha	Badie
Ceiba pentandra	onyina
Celtis philippensis	Esafufuo
Cola caricaefolia	Ananseaya
Cola gigantean	Watapuo
Cordia millenii	Twenwboa
Cumbretum adenogonium	Cumbretum
Cylicodiscus gabunensis	Denya
Danielliia oliveri	Senya
Diospyros abisinica	Dongoro
Diospyrosvirindicans	Atwea
Elaeophorbia grandifoloia	Akan
Entada africana	Entada
Erythrophleum ivorensis	Potrodom
Ficus elasticoides	Amangeydua
Ficus exasperate	Nyankyereni
Ficussur	Adoma
Ficussur	Domini
Funtumia elastic	Funtum
Gardenia ternifolia	Bohyehye
Hildegardia barteri	Abisa

Table 3. (Continued)

Scientific name	Local name
Holarrhena floribunda	Sese
Khaya senegalensis	Mahogany
Kigelia Africana	Nufoo
Lannea kerstingii	Kuntukuru
Lecaniodiscupanioides	Dwindwira
Macaranga barteri	Pambaa
Magaritaria discoidea	Pepea
Milicia excels	Odum
Morinda lucida	konkroma
Morus mesozygia	wonton
Nauclea lactifolia	Nauclealactifolia
Nesogordonia papaverifera	Apro
Newbouldia laevis	Sesemasa
Parkia biglobosa	Sronoo
Petersianthus macrocarpus	Esia
Piliostigma thonningii	piliostigma
Pseudospondias microcarpa	Duakoo
Pteleopsissuberosa	Korkorwa
Pterocarpus erinaceus	Krahyire
Ricinodendron	Wama
Spathodea campanulata	Akuakuaninsuo
Sterculia tragacantha	Sofo
Terminalia superba	Ofram
Termina lisglaucescens	Ongo
Triplochiton scleroxylon	Wawa
Zanthoxylum leprieurii	Okuo

Tree numbers diminished where agricultural activities were high. Apart from the economic benefits, there were other ecological and socio-cultural uses of tree resources by the people of the proposed CREMA area. The appreciation of the program manifested itself in how the members embraced the training and fully participated in the field work.

It is therefore recommended that the process of the CREMA should be expedited to slow down the rate of tree depletion.

References

[1] Ticktin, T. (2004). The ecological implications of harvesting non-timber forest products. *Journal of Applied Ecology* 41:11-21.

[2] Cunningham, A. B. (2001). *Applied Ethnobotany: People, Wild Plant Use and Conservation*. Earthscan Publications Ltd, London and Sterling, VA.

[3] Menges, E. (2000). Population viability analyses in plants: challenges and opportunities. *Trends in Ecology and Evolution*, 15, 51-56.

[4] Ticktin, T., Nantel, P., Ramírez, F., and Johns, T. (2002). Effects of variation on harvest limits for non-timber forest species in Mexico. *Conservation Biology*, 16, 691-705.

[5] Bennett, E. L. (1998). *The natural history of Orang-Utan. Kota Kinabalu*. Sabah, Malaysia: Natural History Publications (Borneo).

[6] White, L. and Edwards, A. eds. (2000). Conservation research in the African ain forests: a technical handbook. New York, Wildlife Conservation Society. 444 pp.

[7] Hawthorne, W. and Jongking, C. (2006). *Woody plants of Western African rainforest. A guide to the forest trees, shrubs and lianes from Senegal to Ghana*. Kew, Royal Botanic Gardens. 1023 pp.

[8] Hossain, M. and Hall, J. B. (1996). *The trees of Mole national park.* Accra, G. I. J. press. 56 pp.

In: Sustainable Utilization
Editors: E. D. Wiafe and F. S. Arku
ISBN: 978-1-63482-896-3

Chapter 4

Adaptive Cluster Sampling for Assessment of Non Timber Forest Products (NTFPs)

Francis Kofi Bih *

College of Technology Education, University of Education, Winneba, Kumasi, Ghana

Abstract

Various studies have been conducted on non-timber forest products (NTFPs) over the years. These studies have however, concentrated more on socio-economic issues of NTFPs than methods of inventory. Forest inventory methods developed for timber have been found to be inefficient for NTFPs because of the special characteristics of NTFPs. Efficient inventory methods are essential to provide quality information on the abundance and distribution of the species for sustainable management. This study contributes to methods of NTFP inventory by comparing the efficiency of systematic sampling method with adaptive cluster sampling method for inventorying 3 NTFP species in off- reserve areas of the Goaso forest district, Ghana. *Calamus deeratus*, raphia palms and bamboo were the 3 NTFP species selected through community and market survey and inventoried in the off-reserve areas of the Goaso forest

* Francis Kofi Bih: College of Technology Education, University of Education, Winneba, P. O. Box 1277, Kumasi, Ghana.

district. Adaptive cluster sampling with a systematic base was designed to inventory the three species.

The results showed that adaptive cluster sampling was about 8 times efficient than systematic sampling for all the 3 species inventoried. However, the density of the species was found to be low and this was attributed to habitat destruction. It was suggested that the methods be tried in a forest reserve where habitat destruction is minimum.

INTRODUCTION

Many studies conducted on NTFPs have concentrated on socio-economic issues of NTFPs (e.g., Peters et al. 1989; de Beer and McDormot 1989; Schwartzman 1989, Padock and De Jong 1989; Falconer 1990; Malhotra et al. 1991; May 1991; Wickens 1991; Richards 1992; Karmann 1998 and Grossmann 2000) than biometric aspects of inventory methods. Many of these and other studies from ethnobotany and anthropology uncovered the importance of the then so called 'minor' forest products and the need for their inclusion in forest management. It is already a known fact that for any natural resource to be managed sustainably, sound knowledge of the ecology, spatial distribution and abundance of the resource are essential (Wong, 2000). Such information could be obtained from a number of sources, including indigenous or local peoples' knowledge as well as formal scientific enquiry through forest inventories (Wong, 2000). However, forest inventories have been focused on timber and little has been done on NTFP inventories. Nevertheless, the intense exploitation of traditional products such as bamboo and rattan in South-East Asia has prompted an increasing interest in resource assessment methodology for these products (e.g., Supardi et al. 1999).

In a study to review NTFP inventories, Wong (2000) realised that the few NTFP inventories which have been "undertaken by foresters have most often adapted or grafted NTFPs onto conventional timber inventory. The choice of sampling designs and plot configurations are therefore determined by the needs of the conventional timber inventory and there is little scope of adapting designs to meet the specific needs of NTFPs" (Wong, 2000: 32). Methods for timber inventory have been found to be inefficient for NTFPs due to special characteristics of NTFPs such as being variable, rare or may occur in clumps. For example, although bamboo was included in Ghana's 2001 multi resource inventory, much bamboo clumps could not be captured because the method used was not suitable to capture enough bamboo clumps for analysis (RMSC,

2001). Inefficient methods of inventory yield unreliable information on NTFP resources, which does not provide sound basis for management.

It is therefore not surprising that overexploitation has characterised most NTFPs exploited on a commercial basis (Teweri, 1999). Wong (2000) therefore suggested the development of methods specific for NTFPs.

NTFP inventories in Ghana have taken place only in the forest reserves. In their studies on NTFPs in southern Ghana, Falconer (1992a and b) and Towson (1995a and b) contended that off-reserve areas provide the bulk of NTFPs in Ghana. However, there has been no work done to quantify the abundance and distribution of off-reserve resource. This chapter contributes to methods of NTFP inventory by comparing the efficiency of systematic sampling method with adaptive cluster sampling method for inventorying selected NTFPs in the Goaso Forest District as a case study of off-reserve NTFP quantification. The chapter would answer the following questions:

- What NTFPs are obtainable off- reserve?
- What is their significance to local people?
- What methods are efficient for their assessment?

NTFP INVENTORY

Forest inventory is the procedure of obtaining information on the quantity and quality of forest resources to assist in their management. Although forests provide a variety of goods and services, timber production had dominated the interest of forest managers and hence has been the main objective of forest inventories (Husch et al., 2003). In western European countries forest management methods to regulate timber production were established far back in the 18th century (Speidel, 1972; Kurth, 1994). Forest management inventories are thus timber oriented and little attention has been given to inventory of NTFPs. However, the institutional rediscovery of the significance of the value of other products such as NTFPs, the formulation of the concept of multiple use of forests and maintenance of biodiversity have recently encouraged the resurgence of interest in NTFPs and hence their sustainable management. NTFPs have also been seen as a means of involving local people, especially forest fringe communities, in forest management through their collaboration. The assumption of NTFPs being a means of involving local people in forest management implies that local people, especially forest fringe communities, would be more interested in forest management if they could derive some benefits through the use of NTFPs from the forest.

This implies also that for continued sustenance of local people's interest in forest management, there should be a sustained supply of NTFPs used by local people. If local people participation is necessary for sustainable management of tropical forests, then the sustainable management of NTFPs would also be necessary for sustainable forest management in the tropics.

For the management of any natural resource to be sustainable, sound knowledge of the ecology, spatial distribution and abundance of the resource is essential. Such information could be obtained from a number of sources including indigenous or local peoples' knowledge as well as formal scientific enquiry through forest inventories (Wong, 2000). These inventories provide information on the resource base on which management decision depends. Thus, the suitability of the method of inventory is essential in providing any useful information about the resource. However, in many situations of tropical forest management, timber has been the main focus and little has been done on NTFPs management.

Although there is often considerable indigenous knowledge for specific NTFPs as well as many anthropological and ethnological studies, formal resource assessment of NTFPs, especially in tropical countries, is relatively new and has received little attention to date. Some methodologies have however, been developed by individual researchers in response to local circumstances and the peculiarities of the resource under study but these have been found to be methods used for traditional forest inventory (Wong, 2000). NTFP management has been hampered by their multitude and variety, lack of appropriate methods of assessment, organisational and financial constrains (FAO, 2001). The variety of life forms and distribution pattern of NTFPs (being rare or found in clumps) mean that traditional forest inventory techniques for timber cannot be adapted easily for NTFPs because they tend to be inefficient.

For example in a recent multi- resource forest inventory in Ghana, the method of inventory could not capture significant bamboo clumps, even though bamboo was included in the species selected for the inventory (RMSC, 2002). The inefficiency of traditional forest inventory methods for NTFPs has prompted a call for more properly researched NTFP specific sampling designs and the application of novel sampling strategies to NTFPs (Wong, 2000). NTFP inventories in Ghana have been concentrated on the forest reserves. Falconer (1992) first conducted an inventory of NTFPs within forest reserves of the high forest zones of Ghana and strongly emphasised the need to consider NTFPs resource in off-reserve areas. She declared that off-reserve areas, especially fallow areas, provided the bulk of locally important NTFPs.

However, subsequent forest inventories which included NTFPs have also been conducted in the forest reserves (RMSC, 2002). There is thus lack of information on NTFPs resource outside forest reserves.

OBJECTIVES AND SCOPE OF THE CHAPTER

In Ghana as well as other tropical countries, NTFPs play a vital role in the lives of forest fringe communities and even some people in the cities still depend on NTFP processing and marketing for their livelihood.

Besides, NTFPs are a subset of biodiversity that has utility to humans and their sustainable management would also contribute to enriching biodiversity. However, management of NTFPs has been limited and so is management information. The variety of NTFPs, the multiplicity of interest and disciplines involved in NTFP assessment, organisational and financial constrains, no globally or even nationally recognised common terminology, units of measurements and appropriate methods of assessment all contribute to bewilder the assessment of NTFPs and the resources providing them (FAO, 2001). This chapter aims at contributing to management information on NTFPs in the off-reserve areas of the Goaso forest district of Ghana by:

- Testing efficient methods for inventorying the selected NTFPs;
- Making an inventory of selected NTFPs;

The chapter begins by defining the problem and reviews the importance of NTFPs in the economic and socio-cultural sense. It continues by exploring forest inventory methods and their suitability for timber and introduces the problem of the need for inventory methods suitable for NTFPs.

The methodology includes community and market study as well as different sampling methods used for NTFPs. The selection of each of the methods was based on its suitability for the species concerned.

The results compare the efficiency of systematic and adaptive cluster sampling methods for inventorying selected NTFPs as well as the quantities and distribution of the species inventoried. The chapter ends by discussing the results and drawing some conclusions.

Methodology

Selection of Species for the Inventory

The species inventoried were first selected based on interviews conducted in six communities in the Goaso forest district. The communities were Ahwiawia, Gyasikrom, Chief Camp, Asuopere, Keyasi Camp, and Ntesere. These communities were randomly selected for the interviews. The selection was not meant to give a representative data of the study area but to give a fair overview of the species exploited as NTFPs.

The survey took the form of door to door semi structured interviews. Males and females of randomly selected households of each selected community were interviewed. Goaso and Mim markets were also surveyed for traded NTFPs. The aims of the community and market survey were to obtain from local people NTFPs for subsistence use, income generation and to seek local knowledge on the ecology, distribution as well as history of availability of the species in off- reserves. Information on ecology, distribution and history of availability was to help in selecting appropriate methods and design for the assessment of the species. Based on the criteria of subsistence use and income generation, the species in table 1 were selected for the inventory.

The species were then divided according to their distribution. *Calamus deeratus* (rattan), Raphia and *Bambusa vulgaris* (bamboo) were found to be distributed mainly in patches along streams and rivers. Adaptive cluster sampling with a systematic base was designed to inventory the three species.

A systematic base was used to allow comparison of the efficiency of systematic sampling and adaptive cluster sampling methods because systematic sampling is often used in forest inventories. The other species (consisting of trees and shrubs) were found to be scattered in other land cover types. The land cover types were identified as tree fallow, shrub fallow, cocoa farms and annuals. Description of these cover types are as given.

Thus most monocrops (cash crops e.g. cocoa, oil palm) start as mixed cropping. Monocrop of maize and other crops are on a small-scale. When the farmer intends to use the land for cash crop production, e.g. cocoa, cocoa seedlings would then be interplanted with staple crops likeplantain, cocoyam, cassava, maize etc.

The food crops mature earlier and as they are harvested more space is created for the growing cocoa trees till the food crops recede, leaving a monocrop of cocoa, oil palm or fruit trees. Where only food crops are grown the land is left to fallow after all crops are harvested.

Table 1. Species selected for the inventory

Scientific Name	Local Name	Habitat	Use
Calamus deeratus	Demere	Patches around streams and rivers	Baskets, furniture
Raphia hookeri	Adobe	Patches around streams and rivers	Baskets, drying mats
Bambusa vulgaris	Pampuro	Patches around streams and rivers and long roadsides	Construction, handicrafts
Tetrapleura tetraptera	Prekese	Forest, Farmlands, Fallowlands	Spice, medicinal
Ricinodendron heudelotii	Wama	Forest, Farmlands, Fallowlands	Curving, medicinal, oil from seeds
Daniellia ogea / Guibourtia ehie	Hyedua	Forest, Farmlands, Fallowlands	Incense
Christiana Africana	Sesedua	Forest, Farmlands, Fallowlands	Curving
Nauclea diderrichii	Kusia	Forest, Farmlands, Fallowlands	Curving
Spiropetalum heterophyllum	Ahomakyem	Forest, Farmlands, Fallowlands	medicinal
Rauwolfia vomitoria	Kakapenpen	Forest, Farmlands, Fallowlands	medicinal
Alstonia boonei	Nyamedua/Sinuro	Forest, Farmlands, Fallowlands	Medicinal, carving
Kola nitida	Bise	Forest, Farmlands, Fallowlands	Food
Strombosia glaucescens	Afena	Forest, Farmlands, Fallowlands	Construction
Margaritaria *discoidea*	Pepea	Forest, Farmlands, Fallowlands	Construction
Ficus exasperata	Nyankyeren	Forest, Farmlands, Fallowlands	Medicinal, Abrasive cleaners
Spathodea campanulata	Kuokuonisuo	Forest, Farmlands, Fallowlands	Medicinal
Chrysophyllum albidum	Akasaa	Forest, Farmlands, Fallowlands	Fruit

The fallow period however, may differ from individuals but it is believed to have become shorter as agricultural lands have become scarce.

MAJOR LAND COVER TYPES IN OFF-RESERVES

The major land cover types identified in off-reserve areas of the Goaso forest district were classified into the following:

Annuals

Annual croplands consist of lands currently under production food crops or which show clear signs of recently harvested crops. Here, crops such as maize (*Zea mays)*, cassava (*Manihot esculenta*), plantain (*Musa paradisiaca*), and cocoyam (*Xanthosoma sagittifolium)* are mostly grown.

These crops are mostly grown in mixed cropping but they could also be grown as mono crops. Cocoa seedlings may also be part of such a mixed cropping system. It should be noted that these crops are not always strictly annual, for example plantain and cassava can be left on the land for more than one year. However their characteristics in terms of tree cover are similar to real annual crops. In these lands, a few trees are observed because the crops do not cope well with much shade.

Only large or highly important trees whose shade the crops could tolerate are left in the field. According to Amanor (1996) the trees farmers retain in their farms are usually the tree species that have some value to them.

Cocoa Farms

This land cover is characterised by mainly cocoa trees with bigger and taller trees, often forest species, left to provide some shade for the cocoa trees. Fruit trees such as *Citrus sinensis, Persia Americana,* and *Magnifera Indica* are often found in this land cover type.

A minimum of two stories is observed, i.e., cocoa and fruit trees at the lower and bigger shade trees on the upper storey. In general many shade trees may be found in older cocoa farms than new ones. Farmers view was that the newer varieties of cocoa tolerate lesser shade and that could do better with lesser or no shade.

Shrub Fallow

This is the 'young' fallow, it may be in rotation or not.

Few trees are regenerating, some poles might be seen but especially shrubs (e.g. *Chromolaena odorata*) are dense and vigorous. The land may have been under fallow for about 1 to 5 years.

Tree Fallow

Tree fallow has not been used for 5 or more years. Often such fallow lands may be intentionally left for people to collect their forest and wood products or may be forestland in transition or sacred groves.

Tree fallow consist of a dense vegetation of young regenerated trees and a few old bigger trees. Secondary forest or patches of forest lands were included in this group. The classification above was made according to cover types encountered using systematic cluster sampling.

However, the sampling may have missed other cover types such as grasses or bare areas which may not be common in the area. The use of remote sensing technique could provide a more detailed classification of the cover types in the area. Systematic cluster sampling was designed to inventory the tree NTFP species as well as timber species. Before describing the sampling design, I would like to give a brief overview of the sampling methods employed.

Sampling Method

Sampling is usually preferred in forest inventory to complete enumeration due to reasons which have been explained by many authors (e.g., Schreuder et al., 1993, Johnson, 2000).

The sampling method and the design used in an inventory depend on the distribution of the species, the budget and the objectives of the inventory.

This section gives an overview of some sampling methods which have been traditionally used for timber inventories and have been adapted for NTFP inventory as well as some novel sampling methods suggested to be efficient for NTFP inventory.

Systematic Sampling

In systematic sampling, the sample units are spaced at fixed intervals throughout the population. According to Husch et al. (2003), systematic sampling designs are widely used in forest inventory because they have a number of advantages:

1. They provide reliable estimates of population means and totals by spreading sample over the entire population. Thus systematic sampling provides a more representative sample than random sampling of the same size, especially when sampling a complex population. Since the complexity of a population usually increases as population size increases (e.g. sampling larger heterogeneous landscapes), systematic sampling would usually provide a more better relative representativeness of such populations than would a simple random sampling of the same size (Johnson, 2000). Johnson (2000) however, maintains that systematic sampling might not produce a representative sample of a population with periodicity.
2. They are usually faster and cheaper to execute than designs based on random sampling because the choice of sampling units is mechanical, eliminating the need for a random selection process.
3. Travel between successive sampling units is easier since fixed directional bearings are followed, and the resulting travel time is usually less than that required for locating randomly selected units.
4. The size of the population need not be known since units are chosen at a fixed interval after an initial point has been selected;
5. Mapping could be carried out concurrently on the ground since the field party traverses the area in a systematic grid pattern.

In forest inventory work, the systematic distribution of sampling units can be used with fixed area plots or strips, or points and lines in probability proportional to size (PPS) sampling. However, because the selected sampling units are not independent of each other, a systematic sampling is fundamentally different from simple random sampling (Johnson, 2000). Husch et al. (2003) maintains that for the mean of a systematic sample to be unbiased, some form of random selection must be incorporated in the sampling process and that the only randomisation possible is the selection of one of the fixed sets of systematic samples. The set will depend on the selection of the initial sampling unit in the population.

The initial sampling unit can be randomly selected out of the entire population of units or it may be randomly selected from the first *k* units in the population. In either case, once the first unit is selected, all additional units are selected at an interval *k*. They ascertained however, that many forest inventories start a set of systematically distributed sampling units at some easily accessible, arbitrarily chosen point (point specified by the designer of the inventory), assuming negligible bias in the resulting estimate of the mean. According to Johnson (2000), this arbitrary chosen position is in most cases the midpoint of first k sets of sampling units. He illustrated that in most line-plot timber inventories sampling lines are drawn parallel to the tract boundary, a cardinal direction, or so as to cross, at right angles, the major streams and ridges within the tract. The first of these sampling lines is located at half the sampling line spacing from the left-most point on the tract boundary, which is usually the westernmost point. The first plot on that line is located at the point that is half plot interval from either the upper (usually the northern) or the lower (usually the southern) tract boundary. The position of the set of plots is fixed by this process.

Cluster Sampling

In cluster sampling, sampling units consist of a group of smaller units (recording units) located around a centre in a fixed configuration. The group of subplots (clusters) forms the sampling unit (Husch et al. 2003). The clusters can take on a multitude of different configurations, depending on the number of subplots, the distance between units, and the geometric distribution. Husch et al. (2003) has illustrated some of the common cluster configurations. The configuration used in any specific situation depends largely on the nature of the sampling problem and the universe of interest. Multistage sampling can be applied to cluster sampling. When the sampling process is limited to drawing primary clusters (cluster locations selected from a set of locations), it is referred to as simple, one-stage, or single-stage cluster sampling. However, in two-stage cluster sampling a subsample of clusters is selected from primary clusters and this subsample measured. There is no limit to the number of stages that can be used if subsample units are available to form clusters (Johnson, 2000). In either single or two-stage cluster sampling etc., the number of clusters per location and the number of subunits in clusters may be similar or differ. For example probability proportional to size (PPS) sampling could be used at the subunit.

As noted by Johnson (2000) equal-sized clusters provide advantages in some cases but equality of cluster size is not essential. The method of selecting cluster locations may be random or systematic.

Johnson (2000) offers the following reasons that make cluster sampling procedure attractive for an inventory:

- When the area to be inventoried is large and/or sampling units are widely dispersed and/or difficult or impossible to identify or locate single sampling units;
- When time and costs associated with travelling between and location of single sampling units becomes excessive.

Single-stage cluster sampling with equal cluster size is used in this chapter and statistics for single-stage cluster of equal is given by Husch et al. (2003). However, Cochran, (1977) provides procedure for analysis of two-stage cluster sampling and unequal clusters sizes.

Statistical Efficiency of Cluster Sampling

Statistical efficiency of is measured by the magnitude of the standard error of the mean or total and the cluster sampling design must strive to keep this value as small as possible (Johnson, 2000). This can generally be achieved by using sampling units that include within themselves as much of the heterogeneity of the population as possible to minimise the differences between the sampling units. In conventional timber inventory, this can be accomplished by using either large sample plots, or if plot size is to be held constant, by using rectangular rather than circular plots. By covering larger areas, large plots include more diverse conditions (e.g. differing stand densities) than small plots. Rectangular plots include more conditions than similarly sized circular plots because they are less compact and extend laterally across greater distances (Johnson, 2000). According to Johnson (2000) this idea of including population heterogeneity within the clusters and keeping heterogeneity between the clusters at a minimum must be applied to cluster sampling design so that fewer sampling units would be needed to meet an allowable standard error requirement. Intracluster correlation is used to measure the degree of homogeneity within clusters and it is an important statistic to determine the efficiency of cluster sampling or the optimal cluster size (Schabenberger, 1991).

Secondly, the cluster configuration must be such that field operations are kept as simple and convenient as possible. This implies that the units making up a cluster should be in relatively close proximity to one another. This requirement is found to conflict with the earlier condition since the elements are rarely randomly distributed and elements that are close to one another are more likely to be similar than elements that are far from one another (Johnson, 2000). Hence when designing a cluster configuration, trade-offs must be made. Whether statistical efficiency or sampling convenience dominates the trade-off depends on the situation (Johnson, 2000). Calculating autocorrelation between and within clusters could be a tool to help in the trade-off decision.

The next section will describe adaptive cluster sampling as a form of cluster sampling in which the number of sampling units per cluster depends on the observations made.

Adaptive Cluster Sampling

Adaptive cluster sample designs are in marked contrast to conventional sampling designs, in which the probabilities for selecting samples do not depend on any observed population characteristics. Density of rare and clustered populations is difficult to estimate precisely using conventional sampling designs.

For such rare and clustered populations, adaptive cluster sampling has been found to be appealing because survey efficiency could be maximised compared with more conventional designs such as simple random sampling (Thompson and Seber, 1996; Christman, 1997; Brown, 2003). The purpose of adaptive strategies is to take advantage of population characteristics to obtain more precise estimates of the population parameters with a given amount of effort. According to Thompson and Seber (1996) many populations of animals and plants have aggregation tendencies due to such factors as schooling, flocking, dispersal patterns, and environmental patchiness. Often, the location and shape of the aggregations cannot be predicted before a survey, making traditional means of increasing precision such as stratification insufficient. For such populations, adaptive sampling strategies have been found theoretically to provide a way to dramatically increase the effectiveness of sampling effort (Thompson, 1990, 1991).

Subsequently there have been various applications of adaptive cluster sampling in the field of biology, ecology and environmental studies (e.g. smith et al. 1995, Keith 2000, Vasudevan et al. 2001, Smith et al. 2003).

Yue et al. (2004) also applied the adaptive cluster sampling procedure to assess the density of ant nets in the city forest of Freiburg. They found adaptive cluster sampling to be more efficient than simple random sampling. However, few applications have been done in forestry to assess plant populations. For example, Acharya et al. (2000) assessed rare tree species in Nepal using systematic adaptive cluster sampling and found that for clustered species the efficiency for density estimation increased as much as 500% but for unclustered species it decreased by 40%. Besides Acworth (1999) used adaptive cluster sampling to assess the density of *Prunus Africanus* in Cameroon and found it more efficient than random sampling. The relative efficiency of the adaptive cluster procedure to conventional sampling methods has been found to depend on certain factors and these are considered in the next section.

Factors Influencing Efficiency of Adaptive Cluster Sampling

The relative efficiency of adaptive cluster sampling to conventional sampling methods depends on several characteristics of the population, the design, and cost (Thompson and Seber, 1996). The following characteristics tend to increase the efficiency of adaptive cluster sampling relative to simple random sampling:

- The within-network (or within cluster) variance is close to population variance, which occurs when the population is clustered or having aggregation tendencies (Thompson, 1990; Smith et al., 1995; Thompson and Seber, 1996)
- The population is rare. This implies that the number of units in the population is big relative to the number of units satisfying the condition for neighbourhood sampling. Also the study region is big relative to the area in which the animals or plants would be encountered (Thompson and Seber, 1996).
- The expected final sample size is not much larger than the initial sample size. With alarger initial sample, the ratio of the expected final sample size to the initial sample size tends to be smaller. In a situation where units in a population satisfying the condition for neighbourhood sampling are all detached from each other, adaptive cluster sampling would be less efficient relative to simple random sampling. In this situation the within-network variance is zero and adaptive sampling

adds only edge units. These increase the sample size but do not contribute to the precision of the estimators (Thompson and Seber, 1996);

- The cost of observing units in clusters or networks is less than the cost of observing the same number selected at random throughout the study region;
- The cost of observing units not satisfying the condition is less than the cost of observing units satisfying the condition;
- In some cases, cost savings may be achieved by basing the condition additional sampling on an easy-to-observe auxiliary variable so that the variable of interest does not need to be measured on edge units (Thompson and Seber, 1996).

As has been described above the efficiency of adaptive cluster sampling depends on degree of rarity and clustering of the population, which is unknown prior to sampling, and the final sample size is random, which adds uncertainty to survey planning.

However, if and when a population is sufficiently rare and clustered for adaptive cluster sampling to be efficient remain an open question (Smith et al., 2003). This implies that adaptive cluster sampling if misapplied could result in failure to achieve the desired efficiency or a final sample size that exceeds a survey's budget. Excessive final sample size can result when the biological population is not as rare as or is more widely distributed than anticipated resulting in a situation which practitioners of adaptive cluster sampling refer to colloquially as encountering "the cluster from hell" (Smith et al., 2003). Thus the choice of the neighbourhood and critical value is important in adaptive cluster sampling. If the critical value is set too low when sampling a particular population the final sample size will be excessively large because the networks that are formed will span many units (Brown and Manly, 1998). If the critical value is too large there may be no adaptive selection of units if no unit in the initial sample has a value that reaches the critical value.

Similarly, if the neighbourhood definition is too large the final sample size will be excessively large and if it is too small there may be little adaptive selection of units. Brown (2003) maintained that there is no clear definition of how to choose the neighbourhood and critical value.

Many works have thus focussed on limiting the final sample size using different strategies (Brown and Manly, 1998; Thompson and Seber, 1996; Salehi and Seber, 1997). The next section will seek to review some of these strategies.

Strategies for Limiting Sample Size and Dealing with Unexpected Exigencies

- As has been revealed in the previous sections, adaptive cluster sampling tends to be most efficient when the population is rare and clustered and the final sample size tends to be proportionally not too much larger than the initial sample size. Practically however, one could underestimate the abundance and range of the population in setting the condition for extra sampling making a rigorous adherence to the adaptive procedure adding more units than time or cost allow. The exact design unbiasedness of the adaptive sampling estimators depends on inclusion in the sample of the neighbourhood of every sample unit satisfying the condition. If
- Sampling is curtailed prematurely, biases may arise (Thompson and Seber, 1996; Su and Quinn, 2003). The same problem arises in conventional sampling when the entire selected sample cannot be obtained. For example where inaccessibility of an area makes it impossible to measure plots in those areas in a forest inventory and biases arise if the completed part of the sample is not representative of the population as a whole. The following guidelines and suggestions have been offered as ways of limiting the total sampling effort in adaptive cluster sampling:
- Stratification of the study area into a number of strata. The study area is stratified into a number of strata ahead of the survey. The neighbourhood definition and critical values are modified if necessary based on the results in the strata already completed e.g. time spent, abundance observed, variance estimated or other criteria (Thompson and Seber, 1996; Brown, 1996a). Each stratum could have a different design and design unbiased estimators of the stratum population total and of estimation of variance depend on the design used in that stratum. The stratified estimate of the population total is the sum of the individual stratum estimates, and the estimate of its variances is the sum of the individual stratum variance estimates (Thompson and Seber, 1996). Selehi and Seber (1995) also suggested two-staged design where the population is partitioned into primary units. Within a selection of a primary unit, secondary units are selected by adaptive sampling. More primary units can be added to the sample if a large critical value has been used and few units have been adaptively selected. However, Su and Quinn (2003) ascertained that stratification

could supplant the anticipated efficiency that ordinary adaptive cluster sampling was intended to provide;

- Using poststratification. This is where the data is stratified during the estimation stage. The initial conventional probability sample would provide an unbiased estimate of the population mean if no adaptive sample were planned at all. Thus priority is given to the initial sample. It is always desirable for cost purposes to do adaptive sampling in sequence whenever a unit satisfies the condition, with travel beginning generally from one side of the study area to the other. If at some point of the survey it is determined that time remains only to finish the initial sample, then all adaptive sampling should stop at that point and the initial sample should be completed. For estimation purposes the population could be post stratified so that one stratum is represented by the adaptive part of the survey and the other stratum is represented by the part of the population sampled conventionally. An adaptive cluster sampling estimator is used in the first stratum and a conventional estimator in the second. The two estimates, along with the two estimate of variance can be combined as in traditional stratified sampling. In general, if the condition for extra sampling is modified one or more times during the survey then the region should be post stratified, with each stratum representing an area within which a given condition was used. The separate stratum estimates may then be combined as stratified estimates. This procedure would be exactly design unbiased had the strata been delineated ahead of time (Thompson and Seber, 1996);
- Selection by order statistics. Where prior knowledge of the population is too weak for the selection of a suitable condition on which to base the extra sampling, the condition can be based on order statistics. This is where adaptive sampling could be carried out in the vicinity of units with the highest values in the initial sample rather than in the vicinity of every observation exceeding a prespecified level. For example, units with highest values in a group of every five initial units are adaptively sampled. Brown (1996a, b) suggested the surveying of all the initial units in the sample and the critical value is set based on the values obtained e.g. the top 10 percentile is used as the critical value (Thompson, 1994). The disadvantage of this approach is that units in the initial sample that do have a value greater than the critical value have to be revisited to conduct the adaptive selection of neighbouring units. In many ecological surveys, such as sampling in a dense forest,

this may be impractical or costly (Brown, 2003). Su and Quinn (2003) also maintained that having a criterion based on the data may make adaptive sampling more feasible by regulating the number of adaptive units added to the sample. They however noted that this condition now changes the criterion for adaptive sampling with each sample and hence the networks of neighbouring units meeting the criterion vary from sample to sample thus making the efficient Horvitz-Thompson estimator no longer unbiased. However, Thompson, (1996) provides unbiased estimator similar to the less efficient Hansen-Hurwitz estimator;

- Brown (1994) describes a way of controlling sample size in an adaptive sampling in which selection of units for the initial sample continues sequentially until a specified sample size is reached. A unit is selected each time and observed with associated adaptive additions. If the cumulative total number of units in the sample is below a specified sample size limit, the next initial unit is added and associated adaptive additions carried out. This continues till the cumulative total sample size equals or exceeds the sample size limit. The final sample size can then vary only by the cluster associated with the last selected sample unit. However, this strategy of sequential selection of initial sampling units as observed by Thompson and Seber (1996) introduces a bias into the estimates of the population total. Although this method controls the overall sampling effort, the need to sequentially select initial units at random preludes using the most efficient travel path between selected units in the study area.
- 'Partition neighbourhoods'. This can be used to limit the potential size of the networks by partitioning the study area into regions or blocks and use a rule that no cluster can spread into a neighbouring block, hence any adaptive addition to the sample stop at these partition boundaries (Thompson, 1994; Thompson and Seber 1996);
- Neighbourhoods can consist of systematic grids of units rather than contiguous units, so that the resulting networks will be constrained to be spatially sparse (Thompson and Seber 1996);
- Utilizing information from a pilot survey as described by Salehi and Seber (1995).

Thompson and Seber 1996 provide estimators of adaptive cluster sampling method.

Figure 1. Early stage of a typical food crop farm in the study area.

Sampling Designs for the Inventory

The species inventoried by the study were grouped into two according to their distribution. *Calamus deeratus*, raphia and bamboo were found to be distributed in patches around streams and rivers. These species were also found to be clumped species. Adaptive cluster sampling was used for these species because it was found to be the appropriate method in view of the given characteristics of the species. The method was thus designed and tested for its efficiency for inventorying the three species in a pilot study and used in a subsequent main inventory. The tree species used as NTFPs (Table 1) but are scattered on the different cover types were inventoried using systematic cluster sampling. This method was found to be the most desirable due to the large size of the area and the funds available for the inventory. I will describe the two designs in the next sections.

Adaptive Cluster Sampling with a Systematic Base

Adaptive cluster sampling with a systematic base was designed to inventory the three species; bamboo, rattan (*Calamus deeratus*) and raphia palm, distributed in patches around water bodies.

A systematic base was used so that the efficiency of the systematic sampling design could be compared with the adaptive cluster design. The area was stratified using streams and rivers as basis on the assumption that the species to be inventoried are concentrated in patches around streams and rivers and any bias here is negligible.

Two percent of the streams and rivers in the study area were selected randomly using Arcview (see Figure 2). The yellow lines in the figure represent streams and rivers selected.

At a randomly selected starting point on each selected stream or river, 50m x 50m plots were located on both sides of the stream or river. Systematic plots of the same configuration as the first were then laid along the stream on both sides of the starting point at 500m interval. Sampling along each stream was done with a rule that if no observations were found in five consecutive plots, the site is abandoned. A total of 63 systematic plots were observed. Where an observation of any of the species of interest is made, 50m x 50m adaptive adjoining plots are made till a network of plots with zero observation edge units is formed (Figure 3). Plots with green outline are the adaptive plots.

The number of clumps of bamboos, the base area of each clump, number of culms in each clump and the average diameter of culms in each clump were taken. The base area of each bamboo clump was determined by measuring the orthogonal clump base dimensions. The number of culms in each clump was counted and this was used to form a linear regression model to determine whether the clump base area could be an appropriate predictor of the number of culms in a bamboo clump.

Figure 2. Selection of streams and rivers for the inventory.

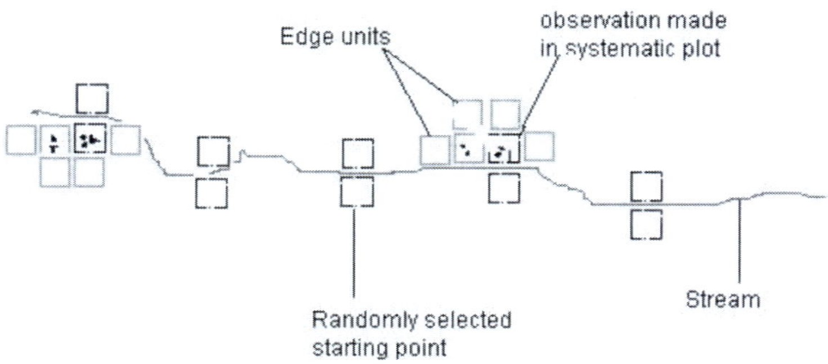

Figure 3. Illustration of the sampling design for the systematic adaptive cluster sampling.

For rattan and raphia palms, the numbers of plants were grouped into mature and immature. In the case of the rattan (*Calamus deeratus*) canes less than 3m length were considered immature while those of 3m length or more were considered mature. With the raphia, palms of less than 3m height were considered immature while those above 3m lengths were considered mature. Estimates were calculated using the Hansen-Hurwitz estimator.

Although the Hansen-Hurwitz estimator is less efficient than the Horvitz-Thompson estimator, they are both superior to simple random sampling (Yue et al., 2004). Since the objective of the study was to compare the efficiency of adaptive cluster sampling with systematic sampling, which is based on simple random sampling, any of the two estimators could be appropriate. The data was analysed using Excel.

RESULTS

This section compares the efficiency of adaptive cluster sampling method with systematic sampling method for inventorying 3 species.

Efficiency of Systematic Sampling Method and Systematic Adaptive Cluster Sampling

Efficiency of a sampling method may be described as the ability of the method to yield relatively lesser variation in the estimates of the population when sampling is repeated.

This may be referred to as statistical efficiency. Statistical efficiency could be measured by the sampling error. Theoretically statistical efficiency could be improved by increasing sampling intensity. Increasing sampling intensity would lead to more time being required for the inventory and hence higher cost and uneconomic.

Therefore a sampling method would be more suitable for inventorying a population if the sampling error and time required for the inventory are relatively lower. I present in this section results of this study comparing both statistical and economic efficiencies of systematic and adaptive cluster sampling for inventorying the 3 NTFP species considered.

Comparison of Efficiency of Systematic Sampling and Adaptive Cluster Sampling for Inventorying Bamboo, *Calamus Deeratus* and Raphia Palms

Sampling error was used as a measure of the efficiency of the sampling methods. It is also an indicator of the precision of the estimates. At a given confidence level (95% in this study) sampling error depicts the probability of estimates deviating from what has been estimated if sampling were repeated a number of times. Thus the lower the sampling error the more reliable is the estimate and vice versa. Table 2 shows estimates and the sampling errors using systematic sampling and adaptive cluster sampling method for inventorying 3 NTFP species.

Table 2 compares the statistical efficiency of systematic sampling with systematic adaptive cluster sampling for inventorying bamboo, Calamus deeratus and raphia palms. Sampling error (E %) was used as the basis of comparison. For all the 3 species adaptive cluster sampling design was about 8 times efficient relative to systematic sampling design. Time used for inventory is a function of the cost of the inventory and hence the importance of time when considering efficiency of inventory methods. Inventory method with a high statistical efficiency but requires relatively much time to execute could be impractical because of budget constraints.

On the other hand if an inventory method that requires relatively less time (cheaper) may not be suitable if it yields unreliable estimates of the population. Therefore a balance should be struck between statistical efficiency and economic efficiency when choosing a method to inventory a population. Whether the balance tilts towards statistical efficiency or economic efficiency depends on the purpose and the budget of the inventory.

Table 2. Comparison of efficiency of systematic sample design with adaptive cluster sample design

Species	Systematic Sampling			Adaptive Cluster Sampling			Relative efficiency
	Mean/ha	CI	E%	Mean/ha	CI	E%	**E%S / E%A**
Bamboo	2	0.4-3.3	76.8	2.8	2.5-3.0	8.75	8.77
Calamus deeratus	10.5	0-22.6	114.6	11	10-13	12.86	8.91
Raphia	30.8	15-47	51.5	29	28-30	6.35	8.12

(Confidence Interval (CI) is at 95% probability).

Table 3. Analysis of time needed for systematic and adaptive cluster sampling method

Species	Mean time/sys Plot (min)	tD = tA-tS (min)	No. Sys plots tD	% tD	E%S	E%S+D	E%A
Bamboo	55	564	10	16.2	76.8	74.6	8.8
Calamusdeeratus	52	205	4	6.2	114.6	113.8	12.9
Raphia	55	2450	45	71.2	51.5	41	6.4

Table 3 presents the time analysis of using systematic and adaptive cluster sampling to inventory the 3 species. Time for plot measurement includes travel time, plot demarcation and enumeration. Travel time for adaptive plots was assumed to be zero since adaptive plots were contiguous to systematic plots and to each other. tD shows the extra time used for adaptive plots and No. systematic plots tD is the corresponding number of systematic plots that could be sampled with the extra time for adaptive sampling.

Discussion and Conclusion

Efficiency of Systematic Sampling and Systematic Adaptive Cluster Sampling Methods for Inventorying Bamboo, Calamus Deeratus and Raphia

Table 2 shows that the sampling errors of the systematic sampling for the three species were high. This might be accounted for by conversion of the species habitat to agricultural fields which may have caused fragmentation in their distribution and hence making the conventional systematic sample inappropriate for their inventory. The three species also occur in clumps which further worsens the suitability of the systematic sampling method for their inventory. The adaptive cluster sampling method gave relatively low sampling errors for all the three species because the distribution characteristics of the species tend to favour the use of this method. This confirms the assertion of Thompson and Seber (1996) that adaptive sampling method tends to be more suitable for species with clumping distribution characteristics. Table 3 shows the time analysis of the sampling process. Total time per plot included walking/travel time, time for plot demarcation and enumeration. It could be seen from table 3 (%tD) that about 16%, 6.2% and 71% of the total time for systematic method were added to the adaptive sampling method for bamboo, *Calamus deeratus* and Raphia respectively. Although much time was used to measure adaptive plots in the case of raphia, about 71% more than the systematic sampling, a significant reduction of the sampling error was made (51.5% to 6.4%). An error of about 51% using systematic sampling would make the results very unreliable. The extra time used for adaptive sampling (tD) could have been used to sample 10, 4 and 45 systematic plots on the average for bamboo, *Calamus deeratus* and raphia respectively. However, if these extra systematic plots were added to number of systematic plots already sampled, it could have reduced the sampling error (E %) from 76.8%, 114.6% and 51.5% to (E%S+D) 74.6%, 113.8% and 41% for bamboo, *Calamus deeratus* and raphia respectively. Hence notwithstanding the extra time for adaptive sampling even in the case of raphia (about 71% more), significant reduction of the sampling errors were made using adaptive cluster sampling than with systematic sampling for all the three species and could be worth the cost if appreciable sampling efficiency is desired. Given the very high coefficient of variation of the sampling units, it will require enormous number of systematic plots to reduce the sampling error to the level of the adaptive cluster sampling for all the three species considered.

The time for adaptive sampling in the case of raphia was comparatively higher and lowest in the case of *Calamus deeratus*. Raphia palms were found to be distributed in belts and were less clustered resulting in sampling more adaptive plots. It is therefore not surprising that the systematic sampling method was more efficient with sampling error of 51.5% in the case of raphia. This again confirms that the efficiency of adaptive cluster sampling method depends on the density and clustering degree of the species being inventoried as asserted by Talvitie et al. (2006). Bamboo and *Calamus deeratus* were more clumped and hence lesser numbers of adaptive plots were sampled.

The very high relative efficiency of adaptive cluster sampling design over the systematic sampling may give it an edge over systematic sampling for inventorying the three species in off-reserve areas. However, adaptive cluster sampling may have performed better than systematic sampling because the distribution of the species may have been fragmented by farming activities. For one to see the full picture of the relative efficiency of the two methods in inventorying the three species, it is recommended that the methods be tested in the forest reserves where disturbance of the species habitat may be minimal.

The means for the species were very low and hence their population density. The low population density of the species might be due to habitat destruction. Conversion of the species habitats to agriculture fields seems to seriously hamper their regeneration. Even though there is a regulation for farmers to leave some 10m strip around streams and rivers, this is not strictly adhered to. It was realised from the field that farmers would farm into the stream or river so far as the valley is low and flooding could be avoided. When the land is to be used for crops which could thrive in flooded areas, clearing is done into the stream or river destroying the habitats of the species.

The low population density conforms to the assertion of the communities that the species are now scarce, especially rattan, in off-reserves due to farming activities. Although locals use such plant species for one thing or another, much attention is not paid to the preservation of such plants off-reserve.

Figure 4. A converted field with remnants of raphia palms.

These plants may serve as common pool resources and that those who own lands on which the plants are found may not benefit much from the resources. Hence when there is possibility for the land to be used for crop production which the land owners think could give them more benefits, the land is readily converted without taken into consideration the presence of those species which are of importance to members of the community (see Figure 4). Those who may need the resource have no chance to preserve the resource if they are not the owners of the land. However, the species may be protected by land owners if used for commercial purposes. An example is the use of raphia palms for commercial alcohol production in certain areas of Ghana.

References

Acharya, B., Bhattarai, G., de Gier, A., and Stein, A. 2000. Systematic adaptive cluster sampling for the assessment of rare tree species in Nepal. *Forest Ecology and Management* 137: 65-73.

Acworth, J. 1999. *Prunus Africana*. Striving for sustainable and equitable resource management in Cameroon. *Medicinal Plant Conservation* 5, p. 15-18.

Amanor, K. S. 1996. Natural and Cultural Assets and Participatory Forest Management in West Africa. In: *conference Papers Series No. 8. International conference on Natural Assets*, Political Economy Research institute and Centre for Scientific Environment, Philippines.

Brown, J. A. 1996a. *The efficiency of Adaptive cluster sampling*, Ph.D thesis, Otago University.

Brown, J. A. 1996b. The relative efficiency of adaptive cluster sampling for ecological surveys. *Mathematical and Information Sciences Report Series* B:96/08. Department of Statistics, Massey University, Palmerston North.

Brown, J. A. 2003. Designing an efficient adaptive cluster samp le. *Environmental and ecological Statistics* 10, 95-105.

Brown, J. A. and Manly, B. F. J. 1998. Restricted adaptive cluster sampling. *Environmental and ecological Statistics*, 5, 49-63.

Christman, M. C., 1997. Efficiency of some sampling designs for spatially clustered populations. *Envionmetrics*, 8, 145-167.

Cochran, W. G. 1977. *Sampling techniques*. Third edition. John Wiley and Sons. 428 pp.

de Beer, J. H. and McDermott, M. J. 1989: The economic value of non-timber forest products in Southeast Asia. Netherlands Committee for IUCN, Amsterdam, The Netherlands.

Falconer, J. 1990. The major significance of minor forest products: The local use and value of forests in the West African humid forest zone. *Community Forestry Note 6*. FAO, Rome.

Falconer, J. 1992 a. *Non-timber forest products in Southern Ghana. A summary report*. ODA Forestry Series 2. NRI, Chatham. 23 pp.

Falconer, J. 1992 b. *Non-timber forest products in Southern Ghana. Main report*. NRI, Chatham. 244 pp.

FAO 2001. *Resource Assessment of Non-Wood Forest Products, Experience and biometric principles*. Non-Wood Forest Products 13. FAO Rome.

Grossmann, C. M. 2000. *Nicht Holz-Waldprodukte in Ost-Kalimantan, Indonesien: Analyse der waldwirtschaftlichen und sozio-ökonomischen Aspekte Ihrer Bedeutung und ihres Entwicklungspotentials für die Lokalbevölkerung in einemHolz konzessionsgebiet*. Wiedebusch, Hamburg.

Husch, B., Beers, T. W. and Kershaw, J. A. Jr. 2003. *Forest Mensuration*. 4th ed. John Wiley and Sons Inc. Hoboken, New Jersey. 443 pp.

Johnson, E. W. 2000. *Forest sampling desk reference*. CRC Press LLC. 985 pp.

Karmann, M. 1998. *Formen kleinbäuerlicher Miombo-Waldnutzung, am Beispiel von Vier Dörfernim Haneni-Distict, Tansania*. Dissertation. Albert-Ludwigs-Universität Freiburg.

Keith, B. D. A. 2000. Sampling designs, field techniques and analytical methods for systematic plant population surveys. *Ecological Management and Restoration* 1: 125-139.

Kurth, H. 1994. *Forsteinrichtung. Nachaltige Regelung des Waldes*. Deutscher Landwirtschaftsverlag GmbH, Berlin.

Malhotra, K. C., Poffenberger, M., Bhattacharya, A., and Dev, D. 1991. Rapid appraisal methodology trials in Southwest Bengal: assessing natural forest regeneration patterns and non-wood forest product harvesting practice. *Forest, Trees and People Newsletter* 15/16: 18-25.

May, P. H. 1991. Building institutions and markets for non-wood forest products from the Brazilian Amazon. *Unasylva* 42(165): 9-15.

Padoch, C. and de Jong, W. 1989. Production and profit in agroforestry: an example from the Peruvian Amazon. In: *Fragile lands of Latin America: strategies for sustainable Development*. Browder, J. G., (ed.), Westview Press, Boulder, Colorado. p. 102-113.

Peters, C. M., Balick, M. J., Kahn, F., and Anderson, A. 1989. Oligargic forests of economic plants in Amazonia: utilization and conservation of an important tropical resource. *Conservation Biology* 3: 341-349.

Richards, D. C. 1992. Cited in Tewari, D. D. 1999. Income and Employment opportunities and potential of non-timber forest products (NTFPs): A Case Study of Gujarat, India. In: *Journal of Sustainable Forestry*, Vol. 8(2). The Haworth Press, Inc., New York.

Salehi, M. M. and Seber, G. A. F. 1997. Two-stage adaptive cluster sampling. *Biometrics*, 53, 959-970.

Schabenberger, O. 1991. The coefficient of intraclass correlation as a measure for the efficiency of cluster sampling. In: *Forest inventories in Europe with specialreference to statistical methods*. Köhl, M. and Pelz, D. (Eds.), Swiss Federal Institute for Forest, Snow and Landscape Research, Birmendorf, Switzerland. pp. 232-239.

Schreuder, H. T., Gregoire, T. G. and Wood, G. B. 1993. *Sampling methods for multiresource forest inventory*. Wiley, New York. 446 pp.

Schwartzman, S. 1989. Extractive reserves: The rubber tappers' strategy for sustainable use of the Amazon forest. In: *Fragile lands of Latin America: strategies for sustainable management* (J. O. Browder, Ed.). Westview Press, Inc., Boulder. p. 150-165.

Smith, D. R., Conroy, M. J. and Brakhage, D. H. 1995. Efficiency of adaptive cluster sampling for estimating density of wintering waterfowl. *Biometrics* 51, 777-788.

Smith, D. R., R. F. Villella and D. P. Lemarie. 2003. Application of adaptive cluster sampling to low-density populations of freshwater mussels. *Environmental and Ecological Statistics* 10: 7-15.

Speidel, G. 1972. Planung Im Forstnetrieb. Grundlagen und Methoden der Forsteinrichtung. Verlag Paul Parey. 267 pp.

Su, Z. and Quinn, T. 2003. Estimator bias and efficiency for adaptive cluster sampling with order statistics and a stopping rule. *Environmental and Ecological Statistics* 10, 17-41.

Talvitie, M., Leino, O. and Holopainen, M. 2006. Inventory of sparse forest populations using adaptive cluster sampling. *Silva Fennica* 40(1): 101–108.

Tewari, D. D. 1999. Income and Employment Opportunities and Potential of Non-timber forest products (NTFPs): A Case Study of Gujarat, India. In: *Journal of Sustainable Forestry*, Vol. 8(2): 55-76. The Haworth Press, Inc., New York.

Thompson, S. K., 1990: Adaptive cluster sampling. *Journal of American Statistical Association* 85: 1050-1059.

Thompson, S. K., 1991. Stratified adaptive cluster sampling. *Biometrika* 78: 389-397.

Thompson, S. K. 1994. Factors influencing the efficieny of adaptive cluster sampling. *Technical Report 94-0301*, Center for Statistical Ecology and Environmental Statistics, Department of Statistics, Pennsylvania State University, University Park.

Thompson, S. K. 1996. Adaptive cluster sampling based on order statistics. *Environmentrics*, 7, 123-133.

Thompson, S. K. and Seber, G. A. F., 1996. *Adaptive Sampling*. John Wiley and Sons, Inc. New York. 265 pp.

Townson, I. M. 1995 a. *Incomes from non-timber forest products: Patterns of enterprise activity in the forest zone of southern Ghana*. Main report. Oxford Forestry Institute, Oxford, UK. 127 pp.

Townson, I. M. 1995 b. *Incomes from non-timber forest products: Patterns of enterprise activity in the forest zone of southern Ghana. Summary report.* OFI, Oxford. 40 pp.

Vasudevan, K., A. Kumar and Chellam, R. 2001. Structure and composition of rainforest floor amphibian communities in Kalakad-Mundanthurai Tiger Reserve. *Current Science* 80: 406-412.

Wickens, G. E. 1991. Management issues for development of non-timber forest products. *Unasylva* 42(2): 1-8.

Wong, J. L. G., 2000. The biometrics of non-timber forest product resource assessment: a review of current methodology. *Research paper for the European tropical Forest Research Network (ETFRN)*, Department for International Development (DFID), UK.

Yue et al. Yue, C., Pelz, D. R., and Klimetzek, D. 2004. Adaptive Klumpenstichprobenverfahren am Beispiel der Schätzung der Ameisendichte. In: *Deutscher Verband Forstlicher Forschungsanstalten, Sektion forstliche Biometrie und Informatik*, 15 und 16 Jahrestagung, p. 317-325.

In: Sustainable Utilization
Editors: E. D. Wiafe and F. S. Arku
ISBN: 978-1-63482-896-3

Chapter 5

SOLID WASTE COMPOSITION, MATERIAL RECOVERY AND CLIMATE CHANGE. THE CASE OF ACCRA POLYTECHNIC

Richard Amfo-Otu[1,*], John Adu Kumi[2,#] and Foster Kwakye[3,≠]

[1]Department of Environmental and Natural Resources Management, Presbyterian University College Ghana, Akropong-Akuapem, Ghana
[2]Department of Agribusiness, Presbyterian University College Ghana, Akropong-Akuapem, Ghana
[3]Department of Environment and Sanitation, Accra polytechnic, Accra, Ghana

ABSTRACT

It is an undeniable fact that waste management poses huge challenges to city authorities and educational institutions are not

* Corresponding author: Richard Amfo-Otu. Department of Environmental and Natural Resources Management, Presbyterian University College Ghana, Akuapem Campus, P.O. Box 393, Akropong-Akuapem. E-mail: richard.amfo-otu@presbyuniversity.edu.gh.

Adu Kumi: Department of Agribusiness, Presbyterian University College Ghana, Akuapem Campus, P.O. Box 393, Akropong-Akuapem, E-mail: adu-kumi.j@presbyuniversity.edu.gh.

≠ Foster Kwakye: Department of, Accra polytechnic, P. O. Box, E mail: foster2simple@yahoo. com.

exempted. The study examined the solid waste composition, material recovery and implication for climate change in a tertiary institution.

Solid wastes were collected from all the departments for sorting and weighing of components using a spring balance for two weeks. The composition of the solid waste was found as organic food waste (38. 1%), papers (20.1 %), plastic materials (18.3 %), textile (10.0%) metals/cans (7.02 %), glass (4.02%) and inert materials (2.46 %). The composition indicated the potential of material resource recovery to safeguard the exploitation of virgin natural resources. The high decomposable organic components (58.2 %) of the waste implied generation of methane gas into the environment from the disposal site, hence contributed to climate change. The presence of high amount of combustible waste materials (86.5 %) in the waste stream indicated possibility of energy recovery through incineration from the waste. Material resource recovery through source segregation should be considered by the polytechnic to reduce the quantity of waste and cost of waste management as well as encouraging recycling, reuse and recovery.

INTRODUCTION

Solid waste management has created many challenges to city authorities all over the world. The rate of urbanization in recent times is very alarming globally and developing countries have had a fair share of this phenomenon. It has been said that more than half of the world's population now lives in urban centres [24, 28]. It has been predicted that by 2050, the population share of urban dwellers may reach 86 per cent in developed countries and 64 per cent in developing countries [28]. The urban population in Ghana has grown over time and is expected to reach about 63% between 2010 and 2025 [9]. The high rate of urbanisation implies a high demand to change agricultural land to residential and construction areas [27]; it also suggests instigating a higher need for exploitation and destruction of natural resources and ecosystems. The population explosion has been experienced by many parts of the world; change in consumption patterns coupled with economic activities has contributed to high rates of solid waste generation. Degradation of organic fraction of waste under anaerobic conditions mostly in landfills or opened and controlled dumpsites contribute to the production of methane and carbon dioxide which are known greenhouse gases which contribute to climate change [22].

Solid Waste Generation Rate

It has been estimated that the globally solid waste generation is approximately 1.3 billion tonnes per year, and this is expected to increase to about 2.2 billion tonnes per year by 2025 [30]. It is important however to note that, average waste generation rate differ from country to country, regions or continents of the world (Table 1). The developed countries are known to generate more waste than the developing countries, and by extension the rich generate more waste than the poor. Waste generation in sub-Saharan Africa is estimated at 62 million tonnes per year with an average of 0.65 kg/capita/day [30]. Ghana has no accurate data on waste generation but the crude estimates put it at 3 million tons per year [8]. This is a mixed waste with some few material recoveries such as plastics and metals. In Ghana, data on waste generation are mostly estimated for various localities and regions but there is no comprehensive nationwide data (Table 2).

Cost of Solid Waste Management

The rising volume of waste is highly associated with high cost of management to city authorities. In fact, [24] argued that governments in most cases lack the necessary capacity or funding to meet the growing service demand for efficient and effective management of solid waste.

Table 1. Current data on solid waste generation from the Regions of the World

Region	Population (million)	Generation rate (Kg/cap/day)	Quantity of Waste produced
African Region	260	0.65	169,119
East Asia and Pacific region	777	0.95	738,958
Europe and Central Asia region	227	1.1	254,389
Latin America and the Caribbean region	399	1.1	437,545
Middle East and North Africa region	162	1.1	173,545
Organisation for Economic Co-operation and Development	729	2.2	1,566,286
South Asia region	426	0.45	192,410
Total	2,980	1.2	3,532,258

Source: World Bank (2012).

The [30] and [26] indicated that the budget for solid waste management is the single largest for many cities. More so, [15] indicated that solid waste management accounts for 30-50% of municipal operational budgets for collecting only half of the generated waste.

The residual products from economic and social activities of various societies have attracted the attention of many researchers and city authorities in recent times. High volumes of solid waste produced by the global community is very worrying considering the cost of managing such waste, the environmental consequences of such management practices on natural resources conservation and contribution to global climate change processes. Utilization of various products in the urban and rural communities certainly results in the production of solid waste with varied physical and chemical characteristics. The physical and chemical characteristics of waste has been said to be very significant in the discussion of resource utilization and climate change [18].

Solid Waste Composition

The composition of solid waste is one of the important physical properties of waste which helps to discuss the characteristics of solid waste and what material resources can be retrieved to conserve natural resources. It is important to note that every waste material was produced from a natural resource.

Table 2. Estimated Waste Generation for major Regional Capitals in Ghana

Continent	Population	Tonnage per month	Density (Kg/m3)	Kg/Capita/Day
Accra	8,029,200	255,556	294	0.63
Kumasi	3,348,700	156,676	290	0.56
Takoradi	307,840	135,391	330	0.51
Tamale	1,458,900	114,433	320	0.58
Cape coast	1,053,900	117,825	300	0.60
Koforidua	249,00	24,242	340	0.48
Suyani	509,500	84,137	310	0.53

Source: Environmental Health and Sanitation Directorate (2010).

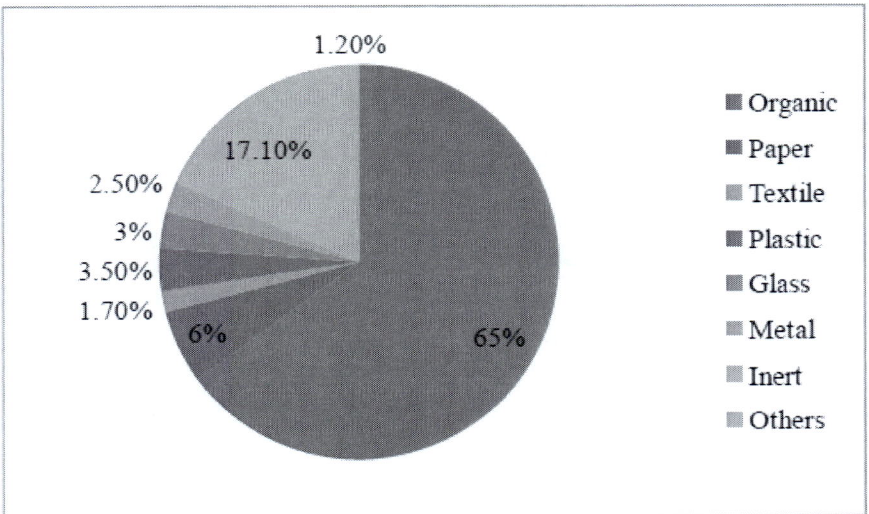

Source: AMA (1999).

Figure 1. Solid Waste Composition for AMA, Ghana.

The waste composition has been defined as the individual material/ elements/components stream that makes up the entire waste and their relative distribution in percentage by weight [2]. The weight is a good measure for the composition because if the waste is compacted there will be no lost in weight of components. [14] and [2] argued that waste composition has the ability to help in developing waste reduction segregation, recycling and resource recovery programmes.

In solid waste operational/collection activities, the composition helps in selecting waste storage facilities; sizing of storage and treatment facilities, determine collection methods and frequency and influences the choice of treatment method. The components that form the waste stream are normally grouped into organics, plastics, papers, metals/cans, textiles, e-waste and inert materials. The composition is said to be influenced by economic status of households, residential class, and economic activities of the area, the season, culture of the people and geographical location. Solid waste composition for [1] indicated high organic material content as shown in figure 1.

Solid waste composition analysis is hardly done by many city authorities and waste management companies to inform operational and management strategies. This create data vacuum for planning, research and management decision making. Polytechnics in Ghana are skilled oriented tertiary institution where technology and innovation are expected to emerge to help the industrialization drive of the country. Yet cost of solid waste management remains one of the high recurrent expenditure to management. This can be partly due to the fact that the composition of solid waste produced in the polytechnic is not documented in literature, no diversion is done to allow for value addition and waste minimisation is not practised. Potential resources to be recovered from the waste are all mixed and dumped at the constrained disposal facility of the Metropolis.

Therefore any difficulty at the landfill site creates huge volumes of uncollected waste at the campus with it associated obnoxious odour and unsightly environment. The study therefore tried to examine the waste composition of Accra Polytechnic to ascertain the potential of material resources recovery, contribution to sustainable resource utilisation and it possible implication on climate change.

CONCEPT OF SOLID WASTE

Solid waste has been defined severally by many authors. First of all, [20] looks at waste as “a man-made thing that has no purpose; or is not able to perform with respect to its purpose”. He argued that object should have purpose, and be able to perform that purpose to the owner, and if it is not then it might be irreversibly damaged in structure and therefore cannot serve the purpose at the time. This argument seems to be cogent as it tries to elucidate issues about when an object is said to be waste. However, this concept does not look beyond the physical object and the materials used for the production of the object to allow for resource recovery. Secondly, [13] proposed a definition which consider waste as “either an output with (a negative market) no economic value from an industrial system or any substance or object that has been used for its intended purpose (or served its intended function) by the consumer and will not be reused”. This means that once a product serves its purpose to the owner, it is no longer needed by any other person which is not wholly true. This definition ties the economic value of the object to only the purpose of creating the object and forgets about the after use value and the salvaging value of the raw materials used in creating the product.

Furthermore, [25] defined solid waste as all the waste arising from human and animal activities that are normally solid and are discarded as useless or unwanted. This definition implies that an object become waste once it is discarded and the fact that it is useless without reference to the owner of the object. An object can be a waste indeed to one person but a great asset or resource to another.

In this study, the concept of waste should be seen in more optimistic way "as a material in transition (MINT)" as proposed in the National Environmental Sanitation Strategy and Action Plan [16] of Ghana.

The [16] conceptualise waste as "a material resources which is not to be discarded but value added on at various stages while in transition within the production and consumption cycle".

This conceptual definition of waste is critical for recovery of material resources from waste which will help conserve natural resources through reduced exploitation of natural resources from the environment. In this study, waste is defined as material resource for which the user has no capacity in terms of resources and technology to make productive use of and can pose threat to health and therefore dispose or intend or required by law to dispose of it". The owner of the material can be an individual, a State or an institution.

Solid Waste As a Resource

A major contributor to the global solid waste challenge is the neglect of solid waste as a resource at the wrong place. To reduce the large volumes of waste produced globally, resources or material recovery from the waste is critical. For instance, [29] indicated that resources recovery and value addition to waste are important for reducing waste quantities in the mega cities. The waste hierarchy which policy makers use to make decisions on waste management based on the environmental benefits gives prominence to waste reduction, reuse, recycling and recovery of energy before disposal.

Historically, material resources recovery has existed in the 1950 where aluminium, copper and other metals were recovered and recycled from waste in the United States of America [5] which contributed substantially to waste generation and natural resources exploitation.

Every product produced by and for humanity originates from the use of natural resources extracted from the environment. The disposal of these products after use destroys a lot of resources and over burden the productive and sink functions of the environment.

Predominantly, material recovery has become the preserve of the households with low income in the informal sector. According to [21], city authorities are of the view that formal waste separation schemes to recover resources are expensive therefore they behave passively by relying on the informal sector. He predicted that considering the global population growth and the need to meet demands, material recovery from waste is going to be the way to go to address the myriad needs of mankind.

As indicated by [7] that conservation of natural resources and protection of the environment is a must for waste management to ensure sustainable development.

METHODOLOGY

1. Description of the Study Area

The study was conducted in a tertiary institution called Accra polytechnic with an authority to award Higher National Diplomas (HND), Diplomas, other certificates and degrees in future. This was established by the Polytechnic Law, 1992 (PNDCL 321) which was replaced by the Polytechnic Act (Act 745) 2007. The Polytechnic is located in the heart of the capital city of Ghana, at Tudu. Currently, the institution has a student and staff population of about thirteen thousand nine hundred and fifty (13,950). This large population generates a lot of solid waste therefore an Environment and Sanitation department has been established to oversee the management of all kinds of waste and janitorial services in the institution. The workforce consists of cleaners and sanitary labourers totaling about 100 in all. Solid waste collection to disposal site has been privatized and this alone is estimated to cost the institution about Forty six thousand and eighty Ghana cedis (GH¢46,080) which is approximately 14,400 US Dollars per year. The master plan for the facilities and infrastructure for the polytechnic is shown in Figure 2.

2. Data Collection Material, Instruments and Methods

The study made use of three different methodologies namely: physical observation, physical measurements and physical analysis. The Materials and instruments used for this study were waste containers for the daily collection of waste from the various departments in the polytechnic, hand gloves for sorting the waste, spring balance as a scale for weighing the waste, containers for weighing, plastic sheets to cover the floor and spread the waste on, plastic bag for collection and sorting of solid wastes, nose mask for covering the nose, overall to cover the body, digital camera to record the research process.

3. Data Collection Procedure

The waste was collected daily from the various departments on campus each morning consistently for about two weeks to a sorting site close to the communal container site on the Campus.

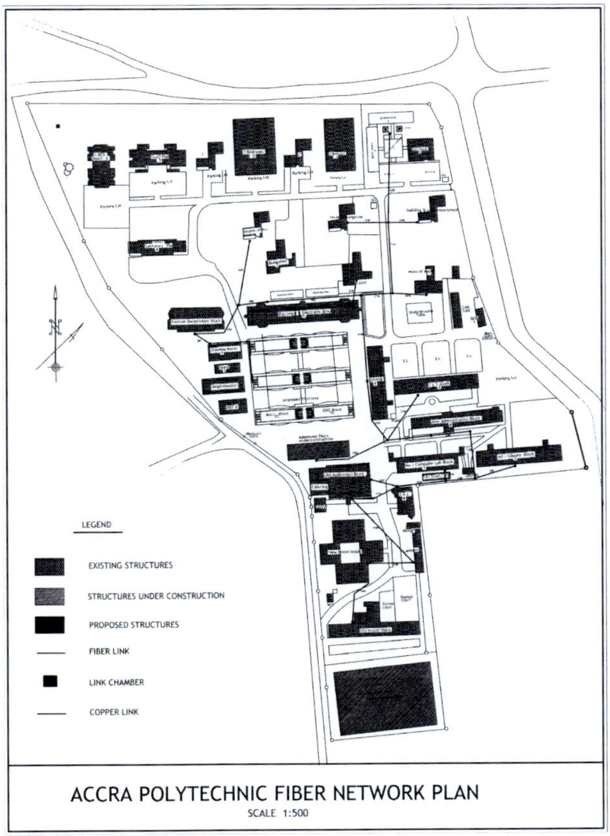

Figure 2. Map of Accra Polytechnic showing the master plan.

Three of the staff of the Environment and Sanitation department were trained to carry out the segregation into identifiable components. The total weight of the waste with the storage container was recorded as well as the weight of the empty container. The weighed wastes were poured unto a spread plastic sheet and sorted into six different categories as: papers, plastics, textiles, metal/cans, glass and organic materials. Each category of waste was weighed separately and results recorded for each day. The data was entered into a spreadsheet for analysis.

4. Waste Composition Determination

Waste storage facility for the blocks was provided and the weight of the storage facility recorded (A). The content of the storage container was weighed and recorded (B) and the waste container was empty every day.

The waste were separated into their respective components, weigh and record each waste component separately (C).

The weight of the empty container was subtracted from the filled container weight (B-A=D)

$$\text{Waste Composition} = \frac{\text{Weight of component}}{\text{total weight of waste}} \text{x100}$$

Figure 3. (Continued).

Figure 3. Sorting Crew Segregating waste into various Components.

RESULTS AND DISCUSSIONS

Results from the waste composition analysis conducted on the campus of Accra polytechnic showed that organic materials content (such as food, yard trimmings, etc.) was the most predominant in the sampled waste stream. The organic waste was 38.10 per cent of the total waste generated in the institution, papers were the second highest (20.1 %), followed by plastic material (18.3 %), the fourth highest percentage was the textile fraction (10.0%) which was followed by metals/cans (7.02 %), followed by glass (4.02%) with the lowest percentage fraction being the inert material (2.46 %) as in figure 3.

The highest component of the waste is organic and this is because most of the waste generated in the polytechnic is food residue. This may also be as result of the fact that 60% of the waste is generated by the polytechnic staff that stays in bungalows, the student hostel and the two major canteens on the

Campus. With all the structures present and fully in use the most likely waste that may come from these structures is simply food wastes which are mostly putrescible. The high level of organic waste in municipal solid waste has been found by other researchers from various parts of the world. For instance, [17] found that the composition of organic waste (44%) from households in WA was more than other components. In a value chain report of [4] it was reported that the organic waste component in the waste from metropolitan, municipal and district assemblies in Greater Accra (GAMA) was 60%.

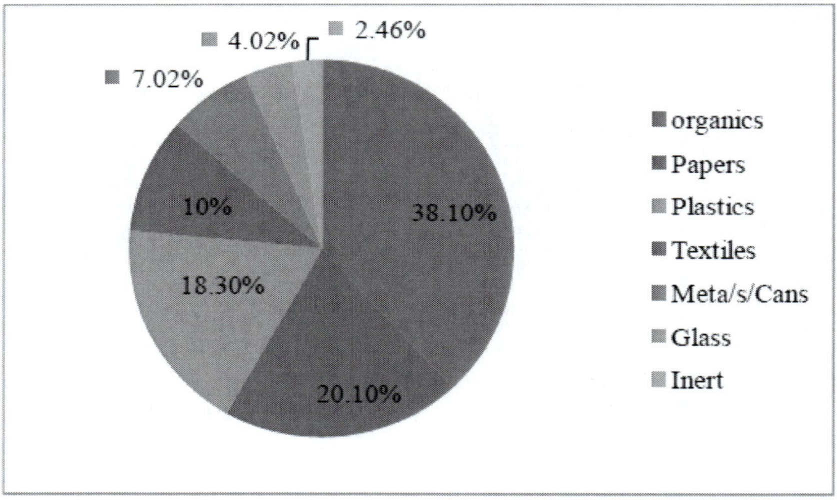

Figure 4. Waste Composition Results for Accra Polytechnic, Ghana.

Even though the amount found for the polytechnic is lower than that reported for the cities, it still reflect the general situation that organic waste component forms the dominant waste generated in Ghana. Furthermore, [11] found from six Secondary Schools in Ogun State, Nigeria, organic waste constituted the largest (62%) of waste produced in the schools.

Even though the amount found for the polytechnic is lower (38.13 %) than what was reported by [11] in Nigeria, the difference may be attributed to the fact that in secondary schools students are fed by the schools by bulk cooking using an existing menu of which food wasting may be a usual characteristics. In polytechnic, students purchase their own food based on preference and therefore food wasting is limited. The amount of organic food waste however confirms [19] who reported 26.8 per cent organic waste component as the highest for the students at the Covenant University, Ota, Nigeria.

It was not surprising that the next highest composition of waste was paper. This can be due to the fact that Accra Polytechnic is an educational institution and for that matter are more likely to produce paper waste since paper is one of their basic tools for studies. This finding contradicts [23] who found paper waste to be more than (29.1%) organic compostable fraction of the waste at the University of Northern British Columbia (UNBC). This gives an indication that at the tertiary institution, paper waste is likely to dominate.

However, [19] found less amount of paper (10.52%) as the fifth component in the waste from the Covenant University, Ota, Nigeria. The next highest component of the waste was plastic waste because plastic waste is tremendously increasing in Ghana and this also shows that Accra polytechnic is not an exception to the plastic waste menace in Ghana. The high amount of plastics is due to the changes in packaging materials used by the production industries and traders.

Others have argued that the organic fraction of waste produced in Ghana is reducing whilst the plastic component is increasing [4, 6]. The Fashion design department contributed highly to the textile component which was the fourth highest.

The percentage of metal in the waste stream is not much because scrape dealers have been scavenging for them leaving only the tin cans to dominate this component. The tin cans are due to the consumption of canned products by both students and staff.

CURRENT WASTE MANAGEMENT PRACTICES AT THE POLYTECHNIC

The waste collection service at the Polytechnic has been privatized. The sanitation workers normally collect the waste from various departments and residential areas on campus to the communal container site. There are two of the 12 m^3 or six tonnes communal containers located at the campus which are lifted twice per week. There is no formal waste separation activities to aid in recovering resources from the waste, therefore all the waste are mixed at the primary and secondary storage points. The private company only collects the accumulated waste at the demarcated container site to the municipal disposal facility. The company does not add any value to the waste but transport all to the disposal site. This implies that valuable material resources are lost or buried at the disposal site. It was found that the Fashion Department of the polytechnic recovered some type of plastic for up-cycling (Figure 1). This act was highly commended because it did not only create job opportunity for people but contribute to the sustainable utilization of natural resources.

RESOURCES RECOVERY POTENTIAL FROM THE WASTE

From the composition analysis it is clear that the waste produced at the polytechnic present huge opportunity for material resource recovery for reuse, recycling, composting and energy recovery. Based on the composition, the waste categories were grouped into two as compostable and combustible material. Compostable materials included organic wastes which were mainly food and papers wastes which could easily be used for composting while combustible materials constituted the combination of organics, papers, plastics and textiles as given by [3]. The high amount of compostable (58.8 %) makes it expedient for the environmental department of the polytechnic to embark on composting to return the nutritional value of the waste into the soil. The organic fraction could be used to generate biogas through anaerobic digestion in digesters. The combustible fractions which constituted the highest fraction (86.5%) could easily be incinerated in appropriate incinerators to generate electricity. This is possible if the calorific value of the combined materials is high to make it efficient system. In fact, [3] indicated that mixing of solid waste components increased the energy potential depending on the colorific values of the combined waste, type and amount of energy-rich bonds in the composite. The recovery of these materials implies the reduction in exploitation of new natural resources which mostly contribute to the degradation of the environment through the destruction of vegetal cover.

Plastic waste takes a long time to degrade in the environment, therefore any effort to use such materials till the fourth or fifth generational use may enhance it environmental degradability. The fashion department in the polytechnic is engaged in the up-cycling of plastics as they occasionally gathered sachet bags and other polyethylene and used them to produce fashioned accessories like hand bags, purse, rain coat, suite, etc. This proactive approach to handle plastic waste materials in the waste helped to reduce the quantity of waste and also to reduce the effect of plastic waste on the environment. It also contributed to job creation and entrepreneurial skills of the students to innovatively use plastic waste products to produce valuable products.

Implication for Climate Change

The decomposition of all the mixed waste with high amount of organic waste at the disposal site is known to generate high amount of methane, carbon dioxide and other trace gases [10].

Even though carbon dioxide from the biogenic sources does not contribute to climate change, the methane gas does and its effect is said to be about 20 to 30 times that of carbon dioxide [12].

This means the current disposal of all waste streams from the polytechnic at the metropolitan disposal site will contribute to climate change and loss of valuable natural resources. Composting of putrescible organic fractions (58.8 %) would have reduced the amount of methane gas (CH_4) produced at the disposal site. This is because the decomposition of organic materials at the landfill site proceeds anaerobically whilst composting is by aerobic processes.

Conclusion

The waste composition for the polytechnic showed the dominance of organic food waste, papers, plastics, textiles, metal/cans, glass and inert materials in the waste stream.

Figure 5. (Continued).

Figure 5. Fashioned Accessories produced from Plastic Waste at Accra Polytechnic.

Material resource recovery from the waste is highly possible to revive the economic value of the material components of products. This would contribute to sustainable utilization of natural resources and reduced environmental degradation.

Processing of compostable components of the waste into compost will help return the nutritional content of the waste into the soil through application in agricultural fields. This will help in reducing the amount of methane gas (CH_4) that would have been produced through anaerobic decomposition at the landfill site. Harnessing the CH_4 through anaerobic processing of the organic waste in digesters would help provide a source of energy and reduce negative environmental effects. The combustible materials present in the waste could be used in generating energy from incineration plants or processed into refuse derived fuel (RDF) for industrial use. The Polytechnic therefore has to integrate a robust and comprehensive source separation of waste scheme into the existing waste management system to ensure proper utilization of the components of the solid waste generated.

REFERENCES

[1] Accra Metropolitan Assembly, (1999). *Annual report: Waste Management Department*. AMA. Accra, Ghana.

[2] Bolaane, B. and Ali, M. (2004). Sampling Household Waste at Source: Lessons Learnt in Cointreau, S. J. (1999). Environmental Management of Urban Solid Wastes in Developing Countries. A project guide. Urban Development Paper. August, No. 5, World Bank, Gaborone, *Waste Management and Research*, Vol. 22, 142–148.

[3] Carboo, D. and Fobil, J. N. (2005). Physico-chemical analysis of municipal solid waste in the Accra Metropolis. *West Africa Journal of Applied Ecology*, vol. 7, 31-39.

[4] CHF International (2010). *Value chain analysis of solid waste management for youth engagement in service delivery (YES) program in Accra*. CHF International/Ghana Accra, Ghana.

[5] Davis, M. L. and Masten, S. J. (2004). *Principles of Environmental Engineering and Science*, 1st edition. McGraw-Hill Companies, New York.

[6] Fobil, J. N. (2000). Municipal Solid Waste Characterization for Integrated Management in the Accra Metropolis, (MSc. Thesis.), University of Ghana, Legon, Accra.

[7] El-Haggar, S. M. (2007). *Sustainable industrial design and waste management: Cradle-to-cradle for sustainable development* (p. 424). Oxford: Elsevier/Academic Press.

[8] Ghana Institute of Engineers (2013). ICE Forum On "Accra Waste Management Strategy" Held. Accessed on 30th November, 2014 at http://ghie.org.gh/ghieHome/index.php?option=com_contentandview=articleandid=1313:ice-forum-on-accra-waste-management-strategy-heldandcatid=106:newsghie-newsandItemid=618.

[9] Ghana Statistical Service, (2005). *Population Data Analysis Report*, Vol. 1, Socio-Economic and Demographic Trends Analysis. Ghana.

[10] Hegde, U., Chang, T. C. and Yang, S. S. (2003). Methane and Carbone dioxide emissions from Shan-Chu-Ku landfill site in northern Taiwan. *Chemosphere*.

[11] Ifegbesan, A. (2010). Exploring secondary school students' understanding and practices of waste management in Ogun State, Nigeria. *International Journal of Environmental and Science Education*, 5(2), 201-215.

[12] IPCC (2007). Climate Change 2007: the physical science basis. In: Solomon, S., Qin, D., Manning, M., et al., (eds.) *Contribution of working Group I to the fourth Assessment Report of the Intergovernmental Panel on Climate Change* (IPCC). Cambridge University Press, p. 996.

[13] Lox, F. (1994). Waste management-Life Cycle Analysis of packaging. Final Report. Study Realised by the Consortium Vrije Universiteit Brussel, Vlaamse Instelling voor Technologisch Onderzoek, Belgian Packaging Institute, for the European Commission, DG XI/A/4.

[14] Makende, E. A. (2007). Complementary Organic and Inorganic Fertilizer Application, Influence on Growth and Yield of Cassava and Maize. *Australian Journal of Basic and Applied Science*. 9,187-192.

[15] Medina, M. (2002). Globalization, Development, and Municipal Solid Waste Management in Third World Cities, El Colegio de la Fronera Norte, Tijuana, Mexico.

[16] Ministry of local government and Rural Development [MLGRD] Environmental Health and Sanitation Directorate, (2010). National Environmental Sanitation Strategy and Action Plan. Government of Ghana, Accra.

[17] Monney, I., Tiimub, B. M. and Bagah, H. C. (2013). Characteristics and management of household solid waste in urban areas in Ghana: the case of WA. *Civil and Environmental Research*, 3(9) 10-21. Available at www.iiste.org.

[18] Mor, S., Ravindra, K., De Visscher, A., Dahiyaa, R. P., and Chandra, A. (2006). Municipal Solid Waste Characterization and its Assessment for Potential Methane Generation: A Case Study. *Sci. Total Environ.* 1; 371 (1-3):1-10.

[19] Okeniyi, J. O. and Anwan, E. U. (2012). Solid Wastes Generation in Covenant University, Ota, Nigeria: Characterisation and Implication for Sustainable Waste Management. *J. Mater. Environ. Sci.* 3 (2) 419-424.

[20] Pongrácz, E. (2002). Re-defining the Concepts of Waste and Waste Management: Evolving the Theory of Waste Management. *Doctoral Dissertation*. University of Oulu, Department of Process and Environmental Engineering, Oulu, Finland. Available online at URL: http://herkules.oulu.fi/isbn9514268210/.

[21] Poulsen, T. G. (2014). Future informal waste material recycling: implementation of return for fees? *Waste management and Research*, 32 (10), 937-938.

[22] Secretariat of the Pacific Regional Environnment Programme (SPREP) (2009). Waste & Climate Change: Facts Sheets. www.sprep.org

[23] Smyth, D. P., Fredeen, A. L. and Booth, A. L. (2010). Reducing solid waste in higher education: The first step towards1 'greening' a university campus. *Resources, Conservation and Recycling*, 54, 1007–1016.

[24] Tacoli, C. (2012). *Urbanization, Gender and Urban Poverty: Paid Work and Unpaid Carework in the City*. International Institute for Environment and Development: United Nations Population Fund, London, UK.

[25] Tchobanoglous, G., Theisen, H. and Vigil, S. (1993). *Integrated Solid Waste Management*. McGraw-Hill, New York.

[26] UN-HABITAT, (2010). Collection of Municipal Solid Waste in Developing Countries. United Nations Human Settlements Programme (UN-HABITAT), Nairobi.

[27] UNEP, (2013). Municipal solid waste: Is it garbage or gold? Accessed on 11th November, 2014 at http://na.unep.net/geas/archive/pdfs/GEAS_Oct2013_Waste.pdf.

[28] UNPD, (2012a.) World Urbanization Prospects, the 2011 Revision. Department of Economic and Social Affairs, Population Division, United Nations, New York. Washington DC.

[29] Wilson, D. C. and Velis, C. A. (2014). Cities and waste: current and emerging issues. *Waste Management Research*, 32(9), 797-799.

[30] World Bank, (2012). What a waste? A Global Review of Solid Waste Management. *Urban development series knowledge paper No. 15*. Urban Development and Local Government Unit World Bank, Washington DC.

In: Sustainable Utilization
Editors: E. D. Wiafe and F. S. Arku
ISBN: 978-1-63482-896-3

Chapter 6

TOWARDS ENERGY SUSTAINABILITY IN GHANA: A REVIEW OF THE EMPIRICS ON DETERMINANTS OF ENERGY CONSUMPTION

Paul Adjei Kwakwa[1,*], Solomon Aboagye[2,#] and Hamdiyah Alhassan[3,≠]
[1]Presbyterian University College Ghana, Okwahu Campus, Ghana
[2]Department of Economics, University of Ghana, Legon – Accra, Ghana
[3]Department of Agricultural and Resource Economics, University for Development Studies, Nyankpala Campus, Ghana

ABSTRACT

This paper reviews and assesses empirical studies on the determinants of energy consumption in Ghana. The review shows that generally, different methodological approaches and data sources have been used in various studies. Existing studies at the micro level have

* Corresponding author: Paul Adjei Kwakwa. Presbyterian University College Ghana, Okwahu Campus, Ghana. E-mail: pauladkwa@yahoo.com.
# Solomon Aboagye: Department of Economics, University of Ghana, Legon – Accra, Ghana. E-mail: sbaboagye@gmail.com.
≠ Hamdiyah Alhassan: Department of Agricultural and Resource Economics, University for Development Studies, Nyankpala Campus, Ghana. E-mail: abena567@yahoo.co.uk.

examined the factors that influence the choice of various energy types while macro based studies have commonly looked at the factors influencing electricity consumption. At the micro level, variables such as education, income level, age, occupation/employment status and gender are found to influence energy choice among households. At the macro level, income, urbanization, level of industrialization and policy reforms affect energy consumption. The review also found that studies on energy intensity at both micro and macro levels as well as renewable energy are scanty and thus it will be helpful for future research to consider these aspects.

Keywords: energy, climate change, emissions and Ghana

I. Introduction

Energy is an important commodity and a key mover of economic growth and development [67]. It is needed for basic human needs (i.e. cooking, heating, lighting and other household-based activities) as well as for sustaining and expanding economic activities like agriculture, electricity production, services, transport and industries. Access to energy has a strong impact on creating opportunities for economic development and no country has substantially reduced poverty without enormously increasing its use of energy [67]. For instance, electricity plays an important role in improving levels of human development and quality of modern life as reported in the Human Development report [63]. In the light of this a sustainable energy supply is paramount.

Meanwhile, a report by the Intergovernmental Panel on Climate Change [59] stated that apart from climate change increasing, it has been induced by human activity, especially by the combustion of fossils fuels. Energy-related greenhouse gas (GHG) emissions account for about 70 percent of total emissions such as methane, carbon dioxide and some quantities of nitrous oxide [59]. Climate change due to rising greenhouse gases (GHGs) in the atmosphere is a serious environmental threat and has potential negative effect for poverty alleviation efforts [64]. Research has shown that, the world most poorest and vulnerable people existing conditions are made worse by climate change. The high incidence of food and water insecurity due to droughts, desertification and severe extreme weather events tends to worsen their condition [65, 58, 57, 61] which trapped them in poverty.

Due to the negative impact of fossil fuels (i.e. coal and crude oil) on the natural environment, as it is associated global warming effects and climate change, there have been calls for the need to look for alternative energy sources. An assessment by the IPCC revealed that the use of clean and renewable energy can help mitigate greenhouse gas emissions [59]. In order to mitigate GHGs emissions, the use of fossil fuel for economic activities must be replaced with renewable and cleaner low-carbon energy sources [62]. Renewable and cleaner low-carbon sources of energy have the potential to open up new possibilities for developing countries to achieve higher development and living standards as well as safeguarding energy resources, the environment and human health [64].

Renewable energy obtained from renewable sources such as sun, wind and water are climate and environmental-friendly with close to zero-carbon emissions during energy generation [67]. Clean energy such as natural gas, compressed natural gas and liquefied petroleum gas (LPG) are low-carbon energy which emit less GHG than conventional energy like coal and fuel wood [64]. Renewable and clean energy enhances energy security and reduces the exploitation of natural resources especially in rural areas where people depend on traditional biofuels like fuel wood, dung and agricultural residues. The use of renewable and clean energy can also increase the quality of life of the consumer because, the time spent or working manpower used in collecting fuel wood is reduced significantly, the efficiency of energy services increase and the negative health impacts decline unlike traditional biofuels [66, 67, 68].

Since the importance of energy consumption cannot be overemphasized and yet its consumption affects the level of emissions that contribute to climate change, it has become very necessary to understand the determinants of energy consumption and to draw relevant implications for energy and climate policy. Accordingly, studies have been conducted in many regions and countries using various data and methodology to explore the determinants of energy consumption (see [69, 70, 71]), either for total energy consumption or particular energy type. Developed countries and some developing countries particularly those from the Asian region have done considerable amount of studies in this regard to appreciate what influences the level of cleaner and unclean energy consumption. However, in the Sub Saharan African region where the people are vulnerable to the negative effects of climate change and at the same time have comparatively low energy accessibility, fewer studies exist. In Ghana for instance, such studies gained attention recently from researchers and scientists. In line with the fact that the country is determined to curtail the effects of climate change and ensure energy security [60], it is crucial to know what influences energy consumption since knowing these factors have a key role partly in reducing GHGs and also ensuring readily availability of energy for the populace in the country. With some considerable amount of interest from researchers now on energy consumption in the country, this paper seeks to review the existing studies on the factors that influence energy consumption in Ghana and identify possible areas that need attention for policy implication.

The rest of the paper is organized as follows: section two reviews theoretical arguments on the determinants of energy consumption; section three reviews present empirical studies on determinants of energy consumption in Ghana; and section four concludes the study.

II. Theoretical Review on the Antecedents of Energy Demand

A principal theory explaining micro-level energy demand is the energy ladder hypothesis according to which, income level is the main determinant of energy choice and demand especially at the household and firm levels. The energy ladder hypothesis explains that there is a movement of household from a particular energy type / source to another, usually from traditional energy type to a comparatively improved energy type, as household income rises. This argument is premised on the fact that, all others being constant, higher income levels imply higher purchasing power and the vice versa. As a result, as household income increases, the following effects are likely to occur with regards to energy choice and demand. First and foremost, the increased income enhances households' ability to buy improved or modern energy type which previously could not be afforded by the households due its expensive nature. Thus, as modern fuels become readily available, consumers acquire modern appliances such as cooking electric and gas stoves, cooling fans, radios, television and fridges [1] and hence moves up the energy ladder.

The implication is that the ownership of appliances and ability to use new fuel type is highly correlated to income [2]. Secondly, the energy types that households need to meet their basic economic needs of cooking, lighting, washing and entertainment changes with income levels. Thus, following the energy ladder model, household substitute comparatively efficient or modern energy type for old energy type by changing the gadgets and appliances used and/ or by changing the type of energy it uses to fuel the old gadgets and appliances. In either choice, the level of income of the household is considered the most crucial.

In particular, the energy ladder hypothesis predicts a three-stage fuel switching and demand processes all of which are closely linked with changes in the income levels of the said economic unit [3, 4, 5, 6]. The first stage is marked by reliance on biomass. In the second stage households move to transition fuels such as kerosene, coal, and charcoal in response to higher incomes and factors such as deforestation and urbanization. In the third phase households switch to liquefied petroleum gas (LPG) and electricity once their income is sufficient [7, 8, 1].

Following the energy ladder model, household decision about the type and quantum of energy demand is captured by several other factors aside income. For instance, Kebede et al. [9] revealed that the high cost of the modern fuels inhibits their usage among low-income households because of their low purchasing power while Campbell et al. [10] concluded that wealthier households are less likely to be affected negatively by the increasingly high cost of non-biomass fuels and they are highly susceptible to make the transition to more sophisticated alternatives. Therefore quite clearly, in the context of increasing energy prices households are confronted with the decision of substituting different energy sources depending on their relative affordability.

Another household characteristic that affect energy demand is the gender of the head of the household. Households headed by females according to Israel [11] have stronger preferences for using a cleaner energy since they are involved a lot in cooking [12]. Thus female headed homes are more likely to have much cleaner energy type than male headed homes. In terms of age, it is argued by Démurger [13] that older people tend to have preference for traditional energy source because they are more conversant to them than the modern ones. Similarly household size, education level, occupation/employment status, and number of income earners in the household are identified to affect energy consumption in the literature [56].

In an attempt to better understand the macro determinants of energy demand among countries and/or across time at the theoretical level, various theories have been proposed by researchers based on different assumptions. Nonetheless, the dematerialization hypothesis remains one of the most widely explored theories explaining energy consumption at the macro level (see [14, 15]). According to this hypothesis there is a substantial rise in material (energy consumption) along the initial path of economic growth but at higher levels of economic growth, technological advancement, climate and environmental issues among other things are expected to cause reduction in materials and hence energy consumption. The dematerialization hypothesis is similar to the prediction of the Environmental Kuznets Curve (EKC).

Although the EKC hypothesis prediction is fundamentally about the existence of an inverted-U shaped relationship between economic growth and environmental degradation, its extension could be usefully applied to understand the energy demand-economic growth nexus. Ramos-Martin [14] argued that since the EKC hypothesis implies that environmental degradation increases with economic activity up to a turning point after which further increase in income tends to improve environmental quality, then by extension a similar relationship could be established for energy demand given that the EKC hypothesis like the dematerialization hypothesis is largely explained by income/economic growth. Theoretically, the EKC hypothesis (or the dematerialization hypothesis) predicts three main effects: scale effects, composition effects and technology effects and since the EKC hypothesis is assumed to reflect the concept of use/consumption, then it is argued here again that energy use/consumption could also be explained by these same three effects (also see [14, 15].

The scale effect represents an increase in energy consumption (and environmental degradation) as a result of more economic activity while the composition effect denotes structural changes which results in changes in the share of each economic activity out of the total economic activity, from agriculture (usually with low energy demand in most developing countries), to industrial activities (usually higher energy demand), and finally back to a low energy consuming activity as services [16, 14]. The technology effect refers to the efficient energy use per unit of output as a result of higher technology development as income levels of a country grows much higher. The joint result of these three effects implies that energy intensity should be much lower in developed economies while developing economies experience higher energy intensity [15].

A review of literature further reveals that prices of alternative energy resources; structural composition of output of an economy and its macroeconomic and demographic dynamics factors are also major drivers of energy demand. Other drivers are trade and foreign direct investment (FDI), climate, technological innovation, energy policies and some other actions of national, state, and local governments (see [14, 15]. The price of commodity is considered crucial to its demand/consumption. Thus, energy prices have substantial influence on energy demand [17, 18]. Ideally, higher energy prices are accompanied by decline in energy consumption and vice versa. In addition, the general price levels in a country could explain variations in energy consumption overtime given the ultimate effect of inflation on real income. Thus, higher inflation is likely to reduce energy consumption through reduced purchasing power [17, 18, 19, 20, 21].

Urbanization is a concept used to describe the large movement of rural population to urban centers of a country and/or a rise in the rate at which the urban population grows. Sadorsky[22] asserts that urbanization is often associated with such consequences such as increased pressure on existing urban facilities, infrastructure, and a rapid rise in the demand for goods and services. In other to meet the demands of the additional population, there is the need to expand these facilities alongside with increase in production in response to the demand by the increasing urban population, a situation which entails massive energy consumption. However, the increased production may lead to economies of scale and thereby reducing energy demand (see [23, 24, 25, 26]).

Again, industrialization which is the introduction and application of new, modern and sophisticated equipments and techniques to the production of existing and new goods and services [22] is found to influence energy consumption. Usually, industrial activities use more energy than does traditional agriculture implying that industrialization has a positive impact on energy demand [22, 27, 25]. Similarly, research and development (R and D) activities have also contributed to declines in industrial energy demand particularly at the firm level. It is expected that this increase in R and D expenditures will lead to more efficient production processes and, therefore, lower energy demand [18, 28].

Technical and structural changes explain most of the changes in energy demand while technical change is considered probably to be the most important factor explaining energy demand decline in the developed world. Also, trade, which comprises exports and imports of goods and services, and FDI potentially, leads to productivity gains via stronger competition for domestic firms due to the presence of foreign owned firms and rivaling imports. Such productivity gains are likely to influence energy consumption [29, 30, 18, 31].

III. Empirical Studies on Energy Demand in Ghana

Empirical studies that address the factors influencing energy choices and/ or energy demand/consumption in Ghana are regarded very paramount given their stern implications for energy conservation policy especially demand-side policy such as energy conservation policies.

This section reviews these empirical studies that have attempted to examine the determinants of energy demand and choices in Ghana. A survey of literature indicates that the empirical literature on energy demand for Ghana has been from both micro and macro dimensions. The former dimension has largely used primary data and in few instances, secondary data, to examine the determinants of various energy types primarily at the household level while the latter has employed almost only purely secondary data for the same analysis.

Micro-Based Empirical Studies

In their study, Kwakwa et al. [32] noted that the expansion of national electricity gridlines and the subsidization of Liquefied Petroleum Gas (LPG) by Government of Ghana to reduce poverty, and the dependence and usage of fuel wood (firewood and charcoal) are not realizing much in terms of results, a situation which could be attributed in part to the evidence that majority of Ghanaians still rely heavily on fuel wood. The study found from a survey of 507 household heads from the forest and savannah zones of Ghana that, about 74.2% of households use electricity, 72.6% use charcoal, 57% used firewood, 32.5% used kerosene and 36.1% used LPG (multiple responses was allowed in this regard). Households use these varied energy sources for cooking, warming house, cooling the house, lighting, ironing, heating water and washing. A logistic regression further demonstrated that income; rural residence, education and kerosene usage reduces electricity consumption while the converse was found for employment. The evidence established for electricity was broadly the same for kerosene consumption except that family size was found to increase its consumption. Further, income, employment and family size were all found to increase charcoal and firewood usage while the contrary was established for LPG. The survey also noted that the use of cleaner energy is constrained by electric power outages, high electricity bills, shortage and fear of getting the house burnt in the case of LPG usage. The findings generally suggest urgent measures to sustain the forest to support future energy needs of the country.

Mensah and Adu [33] realized that the rate of a deforestation rate of 2% (22,000 hectares) per annum is a worrisome situation given that chunk of Ghanaians depend heavily on biomass energy with only a few using LPG as their primary source of fuel for cooking. Using data on 8,262 households from the Ghana Statistical Service's fifth round of the Ghana Living Standards Survey (GLSS V), they found that about 89.2% of Ghanaians use biomass as a main cooking source and the rest use modern energy. Specifically, 31.9% of households use charcoal, 1.2% rely on crop residue for cooking energy, 9.9% use LPG while as high as 56.1% use firewood, with only 0.6% and 0.3% of households respectively using kerosene and electricity. Expectedly, rural households were found to rely very heavily on firewood and crop residue while their urban counterparts rather use more of charcoal, kerosene, LPG and Electricity. An ordered probit analysis pointed that larger size, older household head and male household heads significantly reduces the household's inclination of using clean and efficient fuels over the inefficient fuels. Higher education, increased income level, increased access to kerosene and firewood and urban residency were also found to increase the propensity of using cleaner energy source.

Manyo-Plange [34] focused on the crucial role of attitudes and behaviours of women, along with other things, in cooking fuel choices in Ghana. Responses from 100 women from Axim suggested that about 63% of women have inclination towards non-biomass fuel usage for cooking, with about 18% voting for LPG as their main fuel. The switching attitudes of these women were found to be influenced largely by institutional, environmental, cultural and economic factors such as prices of alternative energy types, availability of particular energy sources, and relative cleanliness of available energy sources. Interestingly, only 13% showed signs of potential movement up the energy ladder in event of any energy stress and climate uncertainty while 41% of the women strikingly indicated satisfaction using lower fuels on the energy ladder. Also, Karimu [35] found income, household size, education, availability of alternative energy sources and urban dwellers as the main drivers of household choice of modern, transition and solid fuels in Ghana. Furthermore, in a survey of 200 households selected through a multistage sampling in the Wa Municipality of Ghana, Kuunibe et al. [36] found that only 1% and 29% of the households use modern clean fuels of electricity and LPG respectively for cooking purpose while about 60% and 10% respectively rely on charcoal and firewood which are regarded as unclean energy sources. Although the authors were quick to admit that the 60% rely on charcoal because it is cleaner than firewood, a plausible explanation for why about 30% use LPG and electricity was completely missing. Additionally, a very significant relationship was established between the marginal propensity to choose modern fuel and factors such as level of education, household income, household size and the availability of traditional fuel types. A logistic regression showed the level of education, income and relative price reduce the probability a household would use biomass while the size of a household is found to have a positive effect.

Abdul-Hanan [37] also noted that the accessible energy for all households in Ghana campaign has yielded little results in rural Ghana as bright lights in the rural areas remain few and very far apart. In a multi-stage sampling survey of 295 households, the authors examined the drivers of households' choice of modern electricity in the West Mampurisi District of Ghana. The Probit model analysis contrary to widely held beliefs, revealed that cattle ownership used as a proxy for wealth/income is not a key determinant of household energy connectivity with tenancy type, radio, Television and fridge ownership, perception about electricity, duration and cost of use of electricity playing out as significant and positive drivers of households' choice of modern electricity in the West Mampurisi District of Ghana. The authors admitted that government's efforts towards the development of alternative energy sources such as renewable energy or promoting the entry of multiple players into the generation market to reduce the cost of supplying utility power to the rural communities while enabling affordability by every household are imperative to enhance accessibility of all households to energy.

Macro-Based Empirical Studies

Like at the micro level, several enquiries have been mounted at the macro or aggregated level to examine factors influencing energy demand in Ghana. Quite noticeably, the macro based studies have principally focused on electricity consumption than any other energy. Prominent among the macro-based studies that focus largely on demand for energy include [38, 39, 40, 41, 42]. Adom et al. [40] concentrated on the electricity sector in Ghana which they argued has evolved to gain much currency in contemporary growth agenda of Ghana. The authors examined the triggers of the ever-rising demand for electricity consumption observed in Ghana by modeling Ghana's electricity demand in line with [43, 44, 45, 46, 47].

Relying on data mainly from the databases of Ener Data Global Energy, CO2 Data Research Services and African Development Indicators for the 1975-2005 period, an Autoregressive Distributed Lagged (ARDL) Bounds to cointegration approach revealed that real per capita GDP, industry efficiency, structural changes in the economy, and degree of urbanization significantly explain aggregate domestic electricity demand in the long run. In the short-run, real per capita GDP, industry efficiency and degree of urbanization significantly explain aggregate domestic electricity demand in Ghana while no evidence was found to support structural changes in the economy as a trigger of rising energy demand.

In a related study, Adom and Bekoe [41] examined the role of policy regime changes in the analysis of energy demand in Ghana. Exploring annual time series data from the databases of Ener Data Global Energy, CO2 Data Research Services and African Development Indicators over 1971-2008 period; three (3) distinct regressions were run to establish the impact of different policy regimes that have occurred at the pre and post Economic Recovery Program (ERP). In the first estimation they ran the full sample regression (1971-2008) while the second and third run the pre ERP (1971-1983) and the post ERP (1983-2008). The findings show that industry efficiency, industry value added and real per capita GDP significantly affected long-run electricity demand for the full-sample estimate and the policy regimes were found to change the level effects of the triggers of electricity demand as the income elasticity of electricity demand for electricity increases from 0.22 during the pre-ERP to 1.93 at the post-ERP period. Similarly, the output elasticity of 0.20 during the pre-ERP increased to 0.53 during the post-ERP. The urbanization elasticity was -0.34 and 0.66 during the pre and post-ERP respectively while that of industry efficiency respectively was -0.86 and −0.47 during the pre and post-ERP periods. The existence of statistically significant difference in the long run electricity demand elasticity prior to the ERP and afterwards was supported by the Chow breakpoint test.

Ackah and Adu [42] investigated into the drivers of high and rising demand for gasoline in Ghana. The authors noted that the period between 1980 and 2010 has witnessed substantial consumption of gasoline fuel in the country but the authors were quick to admit that this could not be explained by price of gasoline energy and income only. In particular, using annual data from 1971 to 2010 from the databases of the International Energy Agency, the World Development Indicators and UNIDO and employing ARDL bounds cointegration approach, a price inelastic demand for transport gasoline was established while output is found to have an inverse relationship with gasoline consumption. Taste and preferences did not have any systematic statistically significant influence on gasoline demand in Ghana. Similarly, in an investigation to uncover, among others, the determinants of electricity in Ghana by employing structural model and surveying time series data spanning from 1971 to 2011 gathered mainly from the International Energy Agency (IEA) and the World Bank, Ackah et al. [48] found positive income elasticity both in the short and long runs. Adom and Bekoe [38] examined the conditional dynamic forecast of electrical energy consumption requirements in Ghana. This was in attempt to fill an extremely important aspect of the energy (electricity) demand issues that have been neglected in the ongoing discourse by previous studies i.e. forecasting of electricity demand. As a result, the authors practically employed the same variables and from the same source used in [40] but with slight modification in the sample period (i.e. 1975–2008). The ARDL and PAM estimations gave the evidence that industrial output, urbanization and income positively affect electricity demand and they exceed the negative effect of the industrial efficiency not only in the short run but in the long run as well. The results from the dynamic forecast estimation reveal that from 2009 to 2020, if real per capita GDP will grow annually by 8% per annum 2009 to 2020 with urbanization also forecasted to grow by 3.11% from between 2009 and 2015 and 2.8% from 2015 to 2020, with industry value-added forecasted to grow at annual average of 1.9% and industrial electricity intensity forecasted to grow annually at 1.5% then Adom and Bekoe [38] revealed the overall electricity demand to increase to 34,867 GWh by the year 2020.

The findings of above macro-based studies, given their reliance on time-invariant models that assume fixed parameters within the sample, suffer from various limitations [49, 50, 51, 52] taken due cognizance of the limitations, Adom [39] examined electricity demand in Ghana using the Rolling regression model, time- time-variant model which assumes that parameters within the sample are stochastic. Exploring annual time series data, 1971-2008 from the World Bank Development Indicators, Ener Data Global Energy and CO2 Data Research Services, a positive effect of income on electricity consumption was established while energy efficiency and economic structure were found to be negative. In addition, the rolling OLS estimates reveal that the estimated demand equation constantly shifts poles for the 29 rolling periods indicating unstable demand elasticities over the period under study, a situation the author traced to policy changes.

Kwakwa et al. [53] researched focused on the effects of economic growth, industrialization, trade openness and urbanization on total energy consumption in Ghana. A Johansen cointegration test shows a long-run relationship exist among all the variables. An error correction model estimation revealed that trade openness reduces energy consumption, while income and industrialization increases energy consumption in the short run. In the long-run, trade openness and urbanization increased energy consumption while income reduces energy consumption. An error correction term value indicated that approximately 50% of the disequilibria from the previous year's shock converge back to the long-run equilibrium in the current year. The authors used annual time series data spanning from 1971 to 2009 from the WDI.

Since Ghana discovered oil and the government of Ghana is committing huge financial resources into its production and commercializing of the gas aspect, Ackah [54] investigated into the determinants of natural gas in Ghana to reveal some knowledge about the natural gas market in the country for informed policy. The author employed the structural time series model and found that natural gas demand in the country is influenced by both economic and non-economic factors. The author used annual data from 1989 to 2009 sourced from the World Bank and the Ghana Petroleum Authority. Adom and Kwakwa [55] Realizing a gap in the energy consumption studies in Ghana, examined the effects of changing trade structure and changing technical characteristics of the manufacturing on the energy intensity in the country. The conclusion from their study among others was that changing technical characteristics of the manufacturing sector and production mix improved energy efficiency while, energy consumption via export and urbanization significantly increases energy intensity.

CONCLUSION

This study has reviewed empirical studies on the determinants of energy consumption in Ghana. Quite clearly, the micro based studies have revealed various drivers of demand for various types of fuel. In this review, notable determinants of energy choice at the micro level include, but are not limited to education, income level, age, occupation/employment status, gender and location. On the other hand, macro based studies have largely emphasized the factors influencing aggregate electricity demand which includes income, urbanization and policy reforms, and trade. It is obvious that very little is still known about the triggers of demand for energy types such as LPG, firewood and charcoal at the macro level. This situation could be possibly owed to data insufficiency on many of these energy types. Also, other very extremely important aspects of the discourse on energy that much is not known for Ghana at both micro and macro levels are energy intensity and demand for renewable energy. Future studies should focus on these areas since it will invariably offer a very useful policy towards resources management.

REFERENCES

[1] Lean, H. H., Smyth, R., (2010), "CO_2 emissions, electricity consumption and output in ASEAN", *Applied Energy*, 87, 1858-1864.
[2] Davis, M. (1998), "Rural household energy consumption: the effects of access to electricity: evidence from South Africa". *Energy Policy*, 26 (3): 207-217.
[3] Njong, A. M. and Johannes, T. A., (2011), "An analysis of domestic cooking energy choices in Cameroon". *European Journal of Social Sciences*, 20, 2.
[4] Alam, M., J. Sathaye and Barnes, D., (1998), "Urban household energy use in India: efficiency and policy implications", *Energy Policy*, 26, 885-891.
[5] Vermeulen, S. J., Campbell, B. M. and Mangono, J. J., (2000), "Shifting patterns of fuel and wood use by households in rural Zimbabwe", *Energy and Environment*, 11 (2000), 233–254.
[6] Kituyi, E., Marufa, L., Huber, B., Wandiga, S. O., Jumba, O. I., Andreae, M. O., and Helas, G., (2000), "Biofuels consumption rates in Kenya", *Biomass and Bioenergy* 20 (2001): 83–99.
[7] Barnes, D. F. and Qian, U., (2002) "Urban inter-fuel substitution, energy use and equity in developing countries". *World Bank Industry and Energy Department Working Paper, Energy Series Paper 53.* Washington, D.C.
[8] Barnes, D. F. and Floor, W., (1999), "Biomass energy and the poor in the Developing World", *Journal of International Affairs*, 53: 237–59.
[9] Kebede, B., Bekele, A. and Kedir, E., (2002) "Can the urban poor afford modern energy?, The case of Ethiopia", *Energy Policy*, 30: 1029–1045.
[10] Campbell, B. M., Vermeulen, S. J., Mangono, J. J., and Mabugu, R. (2003), "The energy transition in action: urban domestic fuel choices in a changing Zimbabwe", *Energy Policy*, 31: 553-562.
[11] Israel, D. (2002), "Fuel choice in developing countries: evidence from Bolivia", *Economic Development and Cultural Change*, 50: 865–890.

[12] Abebaw, D. (2012), "Household determinants of fuel-wood choice in urban Ethiopia: A case study of Jimma town", *Journal of Developing Areas* 41(1): 117–126.
[13] Démurger, S., Hou, Y. and Yang, W, (2009), "Forest management policies and resource balance in China: An assessment of the current situation", *The Journal of Environment and Development*, 18(1), 17-41.
[14] Ramos-Martin, J. (2001), "Population and Environment", *Journal of Interdisciplinary Studies*, 22(3): 281-313.
[15] Recalde, M., Ramos-Martin, J. (2012), "Going beyond energy intensity to understand the energy metabolism of nations: The case of Argentina", *Energy*, 37 (1): 122-132.
[16] Mielnik, O. and Goldemberg, J., 2000. "Converging to a common pattern of energy use in developing and industrialized countries". *Energy Policy* 28 (8), 503–508.
[17] Fan, Y., Liao, H. and Wei, Y., (2007), "Can market oriented economic reforms contribute to energy efficiency improvement? Evidence from China", *Energy Policy*, 35: 2287-2295.
[18] Fisher-Vanden, K., Jefferson, G. H., Liu, H. M., and Tao, Q., (2004), "What is driving China's decline in energy intensity?" *Resource and Energy Economics* 26: 77–97.
[19] Hang, L. M. and Tu, M. Z., (2007), "The impacts of energy price on energy intensity: evidence from China", *Energy Policy*, 35(5):2978–2988.
[20] Kleijweg, A., Leeuwen, G., Van Huigen, R., and Zeelenberg, K., (1989), "The demand for energy in Dutch manufacturing; a study using panel data of individual firms", 1978–1986. *Research Paper 8906*, Research Institute for Small and Medium-sized Business in the Netherlands, Department of Fundamental Research.
[21] He, C. and Wang, J., (2007), "Energy intensity in light of China's economic transition", *Eurasian Geography and Economics*, 48(4): 439-468.
[22] Sadorsky, P., (2013), "Do urbanization and industrialization affect energy intensity in developing countries?" *Energy Economics*, 37:52–59.
[23] Mishra, V., Smyth, R. and Sharma, S., (2009), "The energy-GDP nexus: evidence from a panel of pacific island countries", *Resource Energy Economics*, 31: 210–220.
[24] Krey, V., van O'Neill, B. C., Ruijven, B., Chaturvedi, V., Daioglou, V., Eom, J., Jiang, L., Nagai, Y., Pachauri, S., and Ren, X., (2012), "Urban

and rural energy use and carbon dioxide emissions in Asia". *Energy Economics*, 34 (S3): S272–S283.

[25] Poumanyvong, P., Kaneko, S., (2010), "Does urbanization lead to less energy use and lower CO2 emissions? A cross-country analysis", *Ecological Economics*, 70, 434–444.

[26] Parikh, J., Shukla, V., (1995), "Urbanization, energy use and greenhouse effects in economic development — results from a cross-national study of developing countries", *Global Environmental Change* 5:87–103.

[27] Samouilidis, J. E. and Mitropoulos, C. S., (1984), "Energy and economic growth in industrialized countries: the case of Greece", *Energy Economics*, 6:191–201.

[28] Garbaccio, R. F., Ho, M. S. and Jorgenson, D. W., (1999), "Why has the energy-output ratio fallen in China?" *The Energy Journal*, 20 (3): 63-92.

[29] Hubler, M., (2009). Energy saving technology diffusion via FDI and Trade: a CGE model of China. *Kiel Working Paper*.

[30] Cole, M. A., (2006), "Does trade liberalization increase national energy use?" *Economics Letters*, 92: 108-112.

[31] Sahu, S. K. and Narayanan, K., (2011), "Determinants of energy intensity in Indian manufacturing industries: A firm level analysis", *Eurasian Journal of Business and Economics*, 4(8): 13-30.

[32] Kwakwa, P. A., Wiafe, E. D. and Alhassan, H., (2013), "Households Energy Choice in Ghana", *Journal of Empirical Economics* 1(3): 96-103.

[33] Mensah, J. T. and Adu, G., (2013), "An empirical analysis of household energy choice in Ghana". Swedish University of Agricultural Sciences, Department of Economics, *Working Paper Series 2013: Uppsala 2013*.

[34] Manyo-Plange, N. C. (2011), "The Changing Climate of Household Energy. Determinants of Cooking Fuel Choice in Domestic Settings in Axim, Ghana". Retrieved from http://www.cleancookstoves.org/resources_files/the-changing-climate-of.pdf.

[35] Karimu, A. (2013), "Cooking fuel preferences among Ghanaian Households: an empirical analysis", Department of Economics, Umeå School of Business and Economics, *Umeå University Doctorial thesis*.

[36] Kuunibe, N., Issahaku, H., Nkegbe, P. K. (2013), "Wood based biomass fuel consumption in the upper west region of Ghana: implications for environmental sustainability", *Journal of Sustainable Development Studies*, 3(2):181-198.

[37] Abdul-Hanan, A., Yeboah, R. W. N., Zakaria, H., and Ibrahim, M., (2014), "What drives household electricity choice? Evidence from

Northern Ghana", *Journal of Agricultural Economics, Extension and Rural Development*: 2(10): pp. 170-179.

[38] Adom, P. K. and Bekoe, W., (2012), "Conditional dynamic forecast of electrical energy consumption requirements in Ghana by 2020: a comparison of ARDL and PAM", *Energy* 44, 367–380.

[39] Adom, P. K., (2013), "Time-varying analysis of aggregate electricity demand in Ghana: a rolling analysis", *OPEC Energy Review* 37, 1, 63-80.

[40] Adom, P. K., Bekoe, W. and Akoena, S. K. K., (2012), "Modeling aggregate domestic electricity demand in Ghana: an autoregressive distributed lag bounds cointegration approach". *Energy Policy* 42, 530–537.

[41] Adom, P. K. and Bekoe, W. (2013), "Modeling electricity demand in Ghana revisited: The role of policy regime changes". *Energy Policy* 61: 42–50.

[42] Ackah, I. and Adu, F., (2013), "Modeling gasoline demand in Ghana: A structural time series", *International Journal of Energy Economics and Policy*, 4(1):76-82.

[43] Houthakker, H. S., (1951), "Some calculations on electricity demand in Great Britain", *Journal of the Royal Statistical Society. Series A (General)*, 114 (3), 359-371.

[44] Houthakker, H. S., Taylor, L. D., (1970), "*Consumer Demand in the US*", 2nd edition. Harvard University Press, Cambridge, MA.

[45] Amusa, H., Amusa, K., Mabugu, R., (2009), "Aggregate demand for electricity in South Africa", *Energy policy*, 37, 4167–4175.

[46] Zuresh, A., Peter, H., (2007), "Electricity demand in Kazakhstan". *Energy policy* 35, 3729–3743.

[47] Lin Bo, Q., (2003), "Electricity demand in the People's Republic of China: investment requirement and environmental impact", Infrastructure Division, East and Central Asia Department of the Asian Development Bank.

[48] Ackah, I., Adu, F. and Takyi, R. O. (2014), "On the demand dynamics of electricity in ghana: do exogenous non-economic variables count?" *International Journal of Energy Economics and Policy*, 4(2):149-153.

[49] Lucas, R. R., (1976), "Econometric policy evaluation: a critique", *Carnegie-Rochester Conference Series on Public Policy* 1, 19–46.

[50] Terasvirta, T. and Anderson, H. M., (1992), "Characterizing nonlinearities in business cycles using smooth transition autoregressive models". *Journal of Applied Econometrics* 7, S119–S136.

[51] Stock, J. H. and Watson, M. W., (1996), "Evidence on structural instability in macroeconomic time series relations", *Journal of Business and Economic Statistics* 14, 11–30.

[52] Phillips, P. C. B., (2001), "Trending time series and macroeconomic activity: some present and future challenges", *Journal of Econometrics*, 100, 21–27.

[53] Kwakwa, P. A. and Aboagye, S. (2014), "Energy consumption in Ghana and the story of economic growth, industrialization, trade openness and urbanization", *Asian Bulletin of Energy Economics and Technology*, 1 (1): 1-5.

[54] Ackah (2014), "Determinants of natural gas demand in Ghana", *OPEC Energy Review*, 38 (3), 272-295.

[55] Adom and Kwakwa (2014), "Effects of changing trade structure and technical characteristics of the manufacturing sector on energy intensity in Ghana", *Renewable and Sustainable Energy Reviews*, 35 (2014) 475–483.

[56] Kwakwa, P. A., Wiafe E. D. and Alhassan, H., (2013), "Household's energy choice in Ghana", *Journal of Empirical Economics*, 3(1): 96-103.

[57] Boko, M., Niang, I., Nyong, A., Vogel, C., Githeko, A., Medany, M., and Yanda, P. (2007), "Africa Climate Change 2007: Impacts, Adaptation and Vulnerability". In: M. Parry, O. Canziani, J. Palutikof, P. van der Linden, and C. Hanson, *Contribution of Working Group II to the Fourth Assessment Report of the Intergovernmental Panel on Climate Change* (pp. 433-467). Cambridge, UK: Cambridge University Press.

[58] Hulme, M., Doherty, R., Ngara, T., New, M., and Lister, D. (2001), "African climate change: 1900-2100. *Climate Research* 17, 145-68.

[59] Intergovernmental Panel on Climate Change (2007), "*Climate Change 2007: Mitigation of Climate Change, Contribution of Working Group III to the Fourth Assessment Report of the Intergovernmental Panel on Climate Change*". Cambridge, United Kingdom: Cambridge University Press.

[60] Ministry of Energy, Ghana (2010). *Energy Sector Strategy and Development Plan.* Retrieved from http://www. ghanaoilwatch.org/ images/energy_strategy.pdf. Accessed on 17th January, 2015.

[61] Morey, J., Milford, L., Madeira, L., and Stori, V. (2011). *Moving Climate Innovation into the 21st Century: Emerging Lessons from other Sectors and Options for a New Climate Innovation Initiative.* Vermont, US: Prepared for the UK Department of International Development and Department of Energy and Climate Change by Clean Energy Group.

[62] Never, B. (2011), "*Integrating Energy and Climate Governance in South Africa: Drivers, Barriers and the Shadow of Hyrarchy*". The Governance of Clean Development Working Paper Series, Working Paper 13. Norwich, UK: University of East Anglia and Economic and Social Research Council.

[63] United Nations Development Programme (2007), "*Human Development Report 2007/2008 Fighting Climate Change: Human Solidarity in a Divided World*", New York: UNDP.

[64] United Nations Economic Commission for Africa (2011), "*Integrating Renewable Energy and Climate Change Policies: Exploring Policy Options for Africa*". UNECA.

[65] World Bank (2009), "Making Development Climate Resilient: A World Bank Strategy for sub-Saharan Africa". *Sustainable Development Department Report.* The World Bank.

[66] World Health Organization (2000), "*Addressing the Links between Indoor Air Pollution, Household Energy and Human Health*", Washington, D.C.: WHO.

[67] World Health Organization (2006), "*Indoor air pollution and health*", Retrieved from http://www.who.int/mediacentre/factsheets/fs292/en/. Accessed on 9th January, 2015.

[68] World Health Organization (2005). *Indoor air pollution. Fuel for life: household energy and health.* Retrieved from http://www.who.int/indoorair/publication/fuelforlife/en/index.html. Accessed on 9th January, 2015.

[69] Sadorsky, P., (2009), "Renewable energy consumption, co_2 emissions and oil prices in the G7 countries", *Energy Economis* 31 (456-62).

[70] Amusa, H., Amusa, K. and Mabugu, R., (2009), "Aggregate demand for electricity in South Africa", *Energy policy*, 37, 4167–4175.

[71] Erdogdu, E., (2007), "Electricity demand analysis using cointegration and ARIMA modelling: a case study of Turkey", *Energy policy*, 35. 1129-114.

INDEX

A

B

C

D

E

F

J

K

L

M

N

S

T

U

V

W

Y

Z